The Eco-physiological and Genetic Basis of Invasiveness

The Eco-physiological and Genetic Basis of Invasiveness

By

Gowher A. Wani and Manzoor A. Shah

Cambridge Scholars Publishing

The Eco-physiological and Genetic Basis of Invasiveness

By Gowher A. Wani and Manzoor A. Shah

This book first published 2020

Cambridge Scholars Publishing

Lady Stephenson Library, Newcastle upon Tyne, NE6 2PA, UK

British Library Cataloguing in Publication Data
A catalogue record for this book is available from the British Library

ISBN (10): 1-5275-5214-4
ISBN (13): 978-1-5275-5214-2

Contents

Introduction

The invasion of ecosystems by alien species is a key driver of global environmental change and many invasive plant species attain sufficiently high abundance to alter ecosystem structure and function (Meffin *et al.* 2010). Thus, attempts have been made to plausibly explain why some alien species undergo a profound shift in their ecological fortune, from being minor components of their native ecosystems to becoming devastating dominants of non-native habitats (Callaway and Maron 2006). To untangle this ecological mystery, the prediction and management of already established and future species invasions undoubtedly holds promise (Hulme, 2003). Research efforts to resolve this paradox have primarily focused on identifying a syndrome of traits that promote invasiveness of alien species in their non-native range(s). Though a multitude of traits contributing to species invasiveness has been recognised, attempts to identify key traits that promote invasiveness of alien species have not yet ceased (Kolar and Lodge, 2001; Pysek and Richardson, 2007). In fact, a number of studies has been carried out to compare the traits of both invasive alien plant species and non-invasive plant species under common environmental conditions (Pysek and Richardson, 2007). Besides conventional morphological, reproductive and physiological traits, recent developments in molecular biology have enabled an exploration of the relationship between the amount of nuclear DNA (genome size) and other molecular attributes and the invasiveness of species (Ekrt *et al.* 2009). Combining the results of these studies in a single meta-analysis enabled Van Kleunen *et al.* (2010) to show that invasive plant species generally have higher values for performance-related traits characterising physiology, leaf-area allocation, shoot allocation, growth rate, size and fitness than non-invasive plant species. This extensive study also pointed out that despite a steady increase, the number of studies comparing traits of invasive and non-invasive species grown under common environmental conditions is still relatively low and very heterogeneous with respect to the type of traits measured and the design of the study. Most studies have included only one or a few species pairs, and each of these studies consequently has a low number of replicates and thus is of limited value. A notable exception, however, is a study of Grotkopp and Rejmanek (2007), which measured seedling growth rates of 15 invasive alien tree species and 13 non-invasive alien tree species. Such experimental multi-species comparisons (also see Van Kleunen *et al.* 2007) considerably increase our power to test which traits are consistently associated with invasive alien species.

Recent studies have showed that invasive species suppress the diversity of native species in their non-native regions but not at home (Shah and Reshi, 2014). The studies so far, though, support the questions of why and how invasive species expand their range in non-native regions, but why these species become rarer in their native regions is still an open question. The advent of new tools in molecular genetics during the past two decades has opened additional avenues to ecologists and biogeographers to address such questions and obtain novel insights into the ecology of invasive species. Molecular tools are especially useful to discriminate between native and exotic populations of invaders that cannot otherwise be differentiated based on morphological traits (Saltonstall, 2002). Population genetics analyses have significantly contributed towards addressing questions of the origin and expansion of invasive species (Sakai *et al.* 2001). Such genetic approaches are being widely used to help control exotic haplotypes and conserve the native species in invaded regions.

Keeping in view the importance of trait analysis studies in the prediction and management of alien invasive species, this book aims to compare the traits of native and invasive aquatic plant species growing in various freshwater ecosystems globally. This focus on aquatic invasive plant species is because of the fact that the biological diversity of freshwater ecosystems is increasingly threatened by the introduction and rapid spread of invasive alien species (Zedler and Kercher, 2004; Olden *et al.* 2006). Recent studies (Johnson *et al.* 2008) have also indicated that the decline and extinction of species in freshwater ecosystems is greater than in terrestrial or marine environments and it is largely due to hydrological alterations and biological invasions. While the focus has been on the study of various facets of plant invasions in terrestrial ecosystems, aquatic habitats have received little attention with respect to the extent and impact of alien invasive species and the traits that promote invasiveness of alien species in these ecosystems (Zedler and Kercher, 2004).

Invasion of Aquatic Ecosystems

Understanding the factors associated with species invasiveness has been a challenging task for ecologists, and given the many interlinked elements of invasion ecology, there is no single structure for a broad review (Richardson and Pysek, 2006). While most of the previous studies regarding the success of invasives have been largely ecological, ecologists have now begun to focus on another potential explanation for the success of invaders: rapid evolutionary change (Bossdorf *et al.* 2005). Considerable research interest has been generated in recent years in developing the predictions of the features of species that enable them to become invasive (Stohlgren and Schnase, 2006; Wani *et al.* 2018).

While humans have greatly facilitated the invasion of aquatic alien species through international shipping and trade, aquarium releases and canal construction etc. (Rahel, 2002), flowing river waters act as natural conduits for the transport of propagules of alien species (Lavoie *et al.* 2003) across geographical and political boundaries. Aquatic habitats, usually being resource-rich, are particularly more vulnerable to the spread of invasive alien species (Zedler and Kercher, 2004) and many aquatic plants are characteristically invasive (Richardson *et al.* 2000). Once established, these invasive species impair ecological processes, such as the level of the water table, fire regimes, soil quality and nutrient cycling, thereby influencing the flow and chemistry of water, blocking drainage pipes, impeding navigation, and hindering commercial and recreational fishing. Invasive macrophytes often have negative effects on aquatic biodiversity, especially fish and macro-invertebrate communities (Keast, 1984; Lyons, 1989); yet some studies demonstrate that they may also be used positively by macro-invertebrates as well as fish (Cyr and Downing, 1988; Chilton, 1990; Bickel and Closs, 2008). Although biological invasions have been implicated as a principal cause of freshwater extinctions (Miller *et al.* 1989; Witte *et al.* 1992), researchers are confronted with the question of whether non-indigenous species drive or follow ecological changes leading to biodiversity loss (Didham *et al.* 2005).

In freshwater systems, invasion studies are relatively difficult due to the interaction of myriad stressors with invasive species. Habitat alteration facilitates invasions and exacerbates their impacts by increasing the connectedness of watersheds, thereby reducing native competitors and predators and increasing habitat homogenisation to the benefit of opportunistic invaders (Moyle and Light, 1996; Scott, 2006; Johnson *et al.* 2008). Impoundment and reservoir construction have been shown to promote the replacement of endemic riverine species by cosmopolitan lentic species that would otherwise be poorly suited to natural river flows (Duncan and Lockwood, 2001). In fact, the interaction of invasion and land use has homogenised freshwater biota across multiple spatial scales (Duncan and Lockwood, 2001).

Invasive species often have a set of traits associated that promote invasiveness such as small, short-lived seeds that can germinate without pre-treatment (e.g. freezing), with short juvenile periods and short intervals between large seed crops, a relatively large size, a longer flowering period, vegetative reproduction and perfect flowers (Crawley, 1987, 1989; Perrins *et al.* 1992). The relative competitive performance of native versus invasive species often depends on environmental conditions (Daehler, 2003). A similarity in climate between native and target regions has long been recognised as a basic requirement for successful invasion (Scott and Panetta, 1993). The establishment and spread of alien plants are sometimes limited by an imperfect climate match (Pysek *et al.* 2003). Fluctuations in the availability of resources can also play an important

role in the invasion process (Davis *et al.* 2000; Davis and Pelsor, 2001). Furthermore, it is expected that invasive processes increase due to global climate change (di Castri, 1990; Sukopp, 1998; Dukes and Mooney, 1999; Pysek *et al.* 2002a), fragmentation of natural ecosystems (Rajmanek and Richardson, 1996; Falinski, 1998; Alpert *et al.* 2000) and growing urbanisation (Klotz, 1990; Sukopp, 1998; McKinney, 2002). Ecosystems with anthropogenic disturbances, such as cities or densely populated areas, contain high numbers of alien species (di Castri, 1989; Falinski, 1998; Pysek *et al.* 1998; Sukopp, 1998; McKinney, 2002), whereas natural or near-natural ecosystems (like natural forests or bogs) display a certain ecological resistance against the introduction of alien species (Elton, 1958; Rejmanek, 1989; Falinski, 1998). Some studies on local scales show that a high species richness of alien plants is coupled with a low species richness of native plants in man-dominated habitats (Mclntyre and Lavorel, 1994; Hoffmann, 1998).

In Kashmir (Himalayas), freshwater ecosystems comprise a sizeable area and support a rich biodiversity. Although several studies have been carried out on macrophytic diversity in different lentic and lotic freshwater bodies of Kashmir (Zutshi, 1975; Kak, 1990; Gopal and Chamanlal, 1991; Gopal and Zutshi, 1998), very little is known about the extent of invasion (Reshi *et al.* 2008). On the other hand, a relatively richer body of literature exists on the widespread invasion of non-native species and their impacts in aquatic systems elsewhere (Gollasch, 2002; Genovesi and Shine, 2004; Gollasch and Nehring, 2006; Lodge *et al.* 2006; Minchin, 2007).

Characteristics of Aquatic Invasive Species

Invasions by exotic species are now happening at an alarming rate (Ricciardi, 2007). The Canadian Wildlife Service defines an invasive species as a non-native species that has moved into a new habitat. These species grow and reproduce so aggressively that they displace some of the original components of the native community (White *et al.* 1993). Their spread is often associated with human activities, which enable them to have a wider range of distribution than unconditional dispersal. Invasive species can be introduced unintentionally (e.g. through inadvertent dispersal via cars or boats) or intentionally (e.g. as ornamental plants or for medicinal purposes). Several investigators have tried to find common characteristics of biological invaders (e.g. Hayes and Barry, 2008; Whitney and Gabler, 2008). A high dispersal rate, high fecundity, fast growth rate, early fruiting, vegetative reproduction and a wide range of tolerance to environmental conditions were all characteristics commonly found in invasive plants (Rejmanek, 1989. However, others have showed that there were no significant differences in the characteristics that separate invasive species from native and non-invasive species (Goodwin *et al.* 1999), and that predictors of invasibility could vary from species to species (Daehler, 2003). Goodwin *et al.* (1999) noted that the size of the native range was a more efficient variable from which to predict invasion success.

The introduction of invasive species into new habitats is the second greatest cause of biodiversity loss worldwide, after habitat degradation (Wilcove *et al.* 1998). Wilcove *et al.* (1998) stated that invasive species are one of the main stressors to native species. Yet, Gurevitch and Padilla (2004) confirmed that there is no concrete evidence to suggest that invasive species can cause extinctions and that it is the combination of invasion with other stressors (e.g. pollution, habitat destruction) that causes extinctions. Ricciardi (2004) suggested that invasive species can cause local extirpations at the population level. Nevertheless, several impacts of invasive species on habitat diversity have been well documented. Because they compete for space and resources, invasive species may negatively affect native species richness, although impacts tend to vary with invaders (Hejda *et al.* 2009). Moreover, hybridisation between native and invasive species was shown to enhance invasiveness in certain plant species (e.g. *Spartina maritima X Spartina alterniflora, Mentha arvensis X Mentha aquatica, Amelanchier humilis X Amelanchier 'clade B'*; Ellstrand and Schienrenbeck, 2000). Invasive species may have considerable social impacts, mostly on human health but also on our appreciation of the environment (Canadian Food Inspection Agency, 2008). The problem is also of particular concern in terms of biodiversity protection. In fact, 44 of the plant species at high risk in Canada are threatened by exotic species (Canadian Food Inspection Agency, 2008).

Attempts at identifying functional characteristics of successful plant invaders were only marginally successful, so other studies have tried to predict invasion success based on the habitat characteristics and community composition. The "efficient use" hypothesis states that invasive species are more efficient than native ones in terms of photosynthetic efficiency and biomass allocation or that they make a better use of resources (Zedler and Kercher, 2004). The "enemy release" hypothesis suggests that exotic species are more successful in their introduced range of distribution because they escape from natural enemies and develop novel competitive abilities in a new environment, unlike native species (Mitchell and Power, 2003; Keane and Crawley, 2002). The "diversity-invasibility" hypothesis proposes that more diverse plant communities are less easily invaded because they offer less availability of resources (Knops *et al.* 1999). However, Van Holle and Simberloff (2005) showed in an experimental design that plant species richness did not affect invasion success at a small scale. Finally, the "fluctuating resource" hypothesis states that fluctuations in resource supply can reduce competitive pressure and invasions are thus facilitated (Davis *et al.* 2000). In a review of literature, Theoharides and Dukes (2007) found that size and distribution of suitable habitat patches, landscape heterogeneity and dispersal corridors are all factors that facilitate the invasion success of exotic species at the landscape level. Principles of island biogeography theory, which describe factors that affect the species diversity of a specific community on islands, may be applied to other systems for a better understanding of the mechanism of biological invasion. Usually, there are more invasive species in small patches than in large patches, and invasive species in large patches are mainly distributed near the edges (Harrison *et al.* 2001). Similarly, at a broader scale, unfragmented landscapes tend to be less invaded than fragmented ones (Parendes and Jones, 2000). This is explained by the fact that the greater the length of contact between a pristine area and adjacent land uses, the greater the propagule pressure. Propagule pressure is a composite term of propagule number and propagule size (e.g. rhizome, abundance of seeds) in an introduction effort, and the number of introductions of invasive species.

Corridors may increase the invasibility of a site by increasing its patchwork connectivity with sources of propagules (Parendes and Jones, 2000). Rhizomes float on dead material (Minchinton *et al.* 2006) and rivers transport seeds, which may later be dispersed through floods (Egertson *et al.* 2004). The density of roads can also be linked with the spread of invasive species (Tyser and Worley, 1992; Lelong *et al.* 2009). Roads may favour invasion by creating new habitats in the roadsides. As they are commonly disturbed by pollutants and dust, clear-cutting, and inputs of de-icing salt, roadsides are harsh environments that can only be colonised by disturbance-adapted species (Tyser and Worley, 1992; Christen and Matlock, 2006). These areas represent open habitats free of vegetation and constitute ideal conditions for the establishment of invasive species (Tyser and Worley, 1992; Christen and Matlock, 2006). Traffic density along roads can also have an impact on the density of invasive species. In fact, deserted roads have been reported to be less invaded than frequently used ones (Parendes and Jones, 2000). The properties of adjacent land uses may also have significant impacts on the invasibility of a given site. The nature of land uses surrounding a given site may affect propagule pressure. Research conducted in 302 reserves in the Czech Republic showed that protected areas surrounded by homogenous or pristine land uses were less invaded than those surrounded by heterogenous lands (Pysek *et al.* 2002).

Wetlands surrounded by farms receive more nutrients which is reported to favour invasion and allows these species to grow more rapidly and form abundant colonies (Bertness *et al.* 2002; Bobbink *et al.* 2006). Moreover, isolated patches of habitat surrounded by farms are more exposed to propagules than patches surrounded by forest which provide a shade tolerance against some exotic communities (Theoharides and Dukes, 2007). An increase in population density around a given territory may also facilitate its invasibility, as more infrastructure is needed to support larger populations (McKinney, 2002). The development of infrastructure (e.g. houses, roads, parking) threatens the local diversity that is not adapted to such turbulence and therefore this favours tolerant invasive species (Hansen *et al*. 2007).

Invasive Plant Species in Wetlands

Wetlands are among the world's most invaded habitats (Bobbink *et al.* 2006). In the Canadian Great Lakes region, wetlands within agricultural and urban landscapes are the areas most exposed to several negative impacts of changes in sedimentation rates, hydrologic cycles (e.g. retention time), nutrient enrichment and soil compaction (Detenbeck *et al.* 1999).

Indian lakes and wetlands with fewer herbivores, an increase in nutrient inputs and greater vegetation removal were all considered as factors for enhancing the invasive capacities of *Phalaris arundinacea, Lythrum salicaria, Typha X glauca, P. australis* and *Myriophyllum spicatum.* The increased content of nitrogen, phosphorus and potassium was associated with a decrease in plant diversity, as it allows certain species to grow taller, shade competitors and produce more litter which leads to the formation of monocultures (Drexler and Bedford, 2002). The occurrence of monocultures of exotic species in aquatic systems was attributed to high concentrations of sediment and water nutrient levels that characterised wetlands located close to farms (Zedler and Kercher, 2004). Wetlands with hydrological disturbances are more susceptible to invasion and are characterised by a downturn in diversity of native species (Zedler and Kercher, 2004).

Attempts to control invasive species can be both costly and time consuming. Blossey (1999) pointed out possible negative impacts on regional flora and the environment that were induced by forms of control of invasive species, especially chemical control. Thus, some authors have recommended that prevention is often the most cost-effective way to control invasive species (Thuiller *et al.* 2005). To be able to efficiently allocate limited resources to prevent or control invasions, managers of protected areas need spatially explicit information about the risks and negative impacts of invasions (Byers, 2002a). Among different foci that risk assessments may have, identifying pathways of introduction is often designated as being crucial because it enables managers to identify potential sources of propagules (Byers, 2002b).

The common reed (*Phragmites australis*) is a cosmopolitan, hydrophytic grass found in wetlands and estuaries throughout the continental United States (Chambers *et al.* 1999). It presents various ploidy levels, with the tetraploid (4x) being the most widely distributed (Lambertini *et al.* 2006). A number of studies have shown the phenotypic and genotypic diversity of this species, allowing it to occupy various environments with different levels of success (Koppitz, 1999; Kuhl *et al.* 1999). *P. australis* is invasive in India and is also present in North America where it is described as a native. Fossilised rhizomes of the species were dated as being 3000 years old, indicating the long-term and native status of the species in North America (Orson, 1999).

More recently, *P. australis* has been described as a cryptic invader in North America, having both native and exotic haplotypes. Genetic studies of nuclear satellite DNA and chloroplast DNA showed that two haplotypes (M and I) were introduced to North America and 11 different haplotypes were native to this region. Haplotype M is European in origin and is considered invasive and problematic in North America. Haplotype I is believed to have originated from south of the United States and from parts of Asia, and is the most widespread haplotype along the Gulf Coast of the United States (Saltonstall, 2002, 2003a). I Meyerson *et al.* (2010 reported evidence of an additional invasion by another introduced haplotype (L), further complicating the description and history of these non-native *Phragmites* in North America. Field experiments have shown that the exotic haplotype M spreads more rapidly than the gulf haplotype, leading to the quick replacement of the latter (Howard *et al.* 2008). Rapid growth of the Eurasian haplotype M is responsible for the marked progression of the species, which is known for its competitive abilities and aggressiveness. It is believed that this haplotype is able to limit the establishment of more slowly growing native wetland plants, as well as change ecosystem functions and diversity (Chambers *et al.* 1999; Zedler and Kercher, 2004). Hence, this haplotype causes serious wetland conservation problems in north-eastern North America (Lavoie *et al.* 2003). In the *Phragmites*-dominant Great Lakes area, this haplotype is a well-established wetland species (Trebitz and Taylor, 2007).

In Quebec, the Eurasian haplotype M was most likely introduced via cargo in the 18th century and seems to have started its spread in Quebec mainly from the Montreal region (Lavoie *et al.* 2003). Looking at the species distribution change over time and space, it is clear that its spread in Quebec generally coincides with the major road construction. Genetic studies, as well as herbarium records, show that it is the spread of exotic haplotype (Delisle *et al.* 2003; Lelong *et al.* 2007; Jodoin *et al.* 2008). Ditches along highways have been shown to serve as propagation systems for invasive wetland plants, especially because they are nutrient-rich environments (Maheu-Giroux and De Blois, 2007). Currently, the northern distributional limit of *P. australis* in Quebec is limited because of recent changes in climate.

Studies performed so far have shown that *P. australis* is capable of becoming a dominant species in multiple types of environment, such as salt marshes (Silliman and Bertness, 2004; Howard *et al.* 2008), freshwater marshes (Meyerson *et al.* 2000; Meadows and Saltonstall, 2009) and oligohaline marshes (King *et al.* 2007; Meadows and Saltonstall, 2009). Many studies deal with the effects of anthropogenic disturbances on reed performance (Havens *et al.* 2003; Alvarez-Cobelas *et al.* 2008; Lelong *et al.* 2009). However, there are only a few studies that deal with a natural freshwater system containing disturbed and undisturbed sites, and exotic and native haplotypes of reed.

INVASION DYNAMICS IN AQUATIC SYSTEMS

The success of an invasion relies on certain factors like the broad tolerance to environmental stresses, release from enemies, hybridisation, efficient use of nutrients and allelopathy. These are some of the possible explanations that provide insights on an invasive species success (Galatowitsch *et al.* 1999; Zedler and Kercher, 2004).

Introduced species are usually less affected by herbivory. When an exotic plant species is introduced from multiple continents, only a limited number of native herbivores will prefer to feed on it while most of the introduced herbivores tend to be generalist species (Liu and Stiling, 2006). Out of 170 species of herbivores listed in Europe that consume *P. australis*, only 26 are present in North America (Tewksbury *et al.* 2002). In addition, there is no clear evidence whether these 26 species of herbivores are specific to the introduced haplotype of *P. australis* in North America. In fact, knowledge of how specific that pests are to *P. australis* in North America is fragmented.

Lepidopterans rhizedra (native), *Phragmataecia castaneae*, *Schoenobius gigantella*, and *Chilo phragmitella* all feed on the rhizomes of *P. australis* (Tewksbury *et al.* 2002). Other species, such as *Platycephala planifrons*, are known to feed on its stem in Europe (Tewksbury *et al.* 2002).

The invasive *P. australis* benefits from human activities and disturbances and takes greater advantage of perturbations (Havens *et al.* 2003; Howard *et al.* 2008; Hershner and Havens, 2008; Meadows and Saltonstall, 2009). A study showed that the exotic reed becomes predominantly abundant in wetlands delimited by at least 15%-developed areas (King *et al.* 2007). In another study, development infrastructure and transportation were thought to be responsible for 92% of the reeds' presence (Havens *et al.* 2003). Another important factor was a positive correlation between the distance to roads and the presence of the reed (Arzandeh and Wang, 2003; Lelong *et al.* 2009), as large road corridors contribute to a longer photoperiod. The common reed also displays a good tolerance to multiple natural stresses, such as high salinity, hydric stress and high pollutant concentrations (Guo *et al.* 2003; Morris *et al.* 2008; Hershner and Havens, 2008). It performs better under low salinity conditions (Silliman and Bertness, 2004), but can also survive in high salinity levels of 35 g/L (Adams and Bate, 1999). Uptake of toxic metals, such as lead, chromium and mercury, is lower in *P. australis* than in *Spartina alterniflora*, which confirms the selective capacity of its roots (Windham *et al.* 2003). It is also capable of invading habitats of variable pH and organic matter content (Haslam, 1972; Alvarez *et al.* 2005; Lambert and Casagrande, 2006). In Europe, the locations of colonies of *P. australis* are not always easily predicted by environmental variables or flooding patterns, but natural perturbations seem to increase reed survival (Alvarez-Cobelas *et al.* 2008).

Nitrogen-rich conditions create optimal conditions for the growth of *P. australis* as it is able to produce more biomass (Bertness *et al.* 2002; Rickey and Anderson, 2004). Nitrogen enrichment can be caused by human disturbances (such as agricultural activity, roads and proximity to residential areas). The combined effects of human disturbances and increased nitrogen availability were able to explain more than 90% of the variation in *P. australis* distribution in Rhode Island (Minchinton and Bertness, 2003). In another study, nitrogen availability alone was able to explain 50% of the variation in *P. australis* abundance. The model explained a variation of 80% when both available nitrogen and salinity were taken into account (Silliman and Bertness, 2004). Nitrogen is the most common limiting nutrient in the case of *P. australis*; its invasion success is maximised where this nutrient is present in higher concentrations, primarily near agricultural fields (Trebitz and Taylor, 2007).

Genetic studies have shown that cross-hybridisation, i.e. hybridisation between individuals belonging to the native and invasive species, contributes more to the spread of the haplotype M in

Quebec than previously thought (Belzile *et al.* 2009). It results in greater genetic diversity and adaptability, which in turn have been found to promote invasiveness of the introduced species. Hybrids are intermediates between parents and can be more adapted to the particular environment as well as more competitive. Moreover, multiple introductions of invasive species increase the genetic diversity of the invader and the likelihood of hybridisation, and are frequently the cause of invasive hybrid progression (Ellstrand and Schierenbeck, 2000). At the species level, a case study of the invasive *Ulmus pumila* showed an unexpectedly high level of hybridisation and even genetic introgression with native tree species *Ulmus rubra* (Zalapa *et al.* 2010).

Intra-specific hybridisation of invasive *P. australis* with a native haplotype has been successful in a greenhouse experiment (Meyerson *et al.* 2000, but only one genetic study, based on field sampling in the Great Lakes area, suggests that hybrids can be found in nature (Paul *et al.* 2010b). This could be due to the fact that seedlings are rarely seen in field studies (Stanton, 2005) or simply do not survive (Shay and Shay, 1986). Furthermore, it would appear that both invasive and native haplotypes could produce intra-specific hybrid seeds, a phenomenon that was not observed in the previously mentioned greenhouse experiment (Paul *et al.* 2010a).

Allelopathic chemicals are released by an organism into the environment that can inhibit the growth of one or more of its neighbours. Examples of such processes are well known (Callaway and Ridenour, 2004); the successful use by introduced plants confirms the importance of the "novel weapon" hypothesis in competition and invasion ecology. For example, introduced *Centaurea maculosa* produces a chemical, racemic catechin that reduces the growth and germination rate of three native plants of North America: *Koeleria micrantha*, *Festuca idahoensis*, and *Arabidopsis thaliana* (Bais *et al.* 2003). Another well-documented example is the case of *Allaria petiolata*. The chemicals released by this plant species reduce plant growth and the survival of seedlings and interfere with mycorrhizal fungi, decreasing the diversity of fungi (Callaway *et al.* 2008). A recent study showed the potential allelopathic capacity of *P. australis* in the rhizosphere. Rhizomes of *P. australis* release gallic acid, interfering with the structural proteins of roots of other plants (Rudrappa *et al.* 2009). This acid is more efficient when transformed into mesoxalic acid due to exposure to UV in an alkaline soil. This could provide an additional tool for the expansion of *P. australis* into freshwater wetlands.

Asexual reproduction plays an important role in the establishment, growth and maintenance of *P. australis*. This species can expand in all directions with the help of runners and rhizomes (Warren *et al.* 2001). The rapid progression of this exotic haplotype forms a dense colony of reed, possibly leading to monocultures (Zedler and Kercher, 2004). These monocultures can be very dense (>180 stems/m^2) and are capable of changing trophic structures and, most probably, hydrological regimes (Haslam, 1972; Chambers *et al.* 1999). In addition, a number of studies have shown that there is a significant interaction between the amount of available nutrients and the removal of cover for other species. When the level of available nutrients is high, it can maximise all growth parameters (biomass, height) and most expansion parameters (biomass of advancing stems and height) (Minchinton and Bertness, 2003). Being one of the highest macrophytes confers this species an advantage in the competition for light. Moreover, a slower rate of decay has been shown to cause an increase in marsh elevation, the slower decomposition rate of cellulose and the filling of small creeks, which ultimately leads to a simplification of the marsh macro-relief (Zedler and Kercher, 2004; Stanton, 2005). Also, the fact that dead stems from the previous year tend to take a long time to decompose contributes to a drier soil that is better suited to the growth of reed. In fact, reed growth is relatively faster when the level of water is low. In the Boucherville islands area, maximum vegetative growth was recorded in the 1999–2002 time period when the mean water level was 5.72 m compared to 1996–1999 when the mean water level was 6.51 m (Hudon *et al.* 2005). The annual progression speed of rhizomes varies in the literature: 0.52 m to 0.91 m (Philipp and Field, 2005), 0.1 m to 0.94 m (Burdick and Konisky, 2003), 1 m (Haslam, 1972), 1.6 m to 2 m (Warren *et al.* 2001), 0.8 m to 1.2 m (Lavoie *et al.* 2003), and runners can go as far as 10 m (Haslam, 1972; Rice *et al.* 2000). Vegetative growth can be very important and this species will take advantage of any canopy opening (Zedler and Kercher, 2004).

Sexual reproduction provides opportunities for further spread, when conditions permit it. However, seed viability is low (3–7%) for the exotic haplotype (Gervais *et al.* 1993; Maheu-Giroux and De Blois, 2007) and very few studies have reported seedlings in Canada. This could be attributed to the effect of high salinity, shading by other plants (Havens *et al.* 2003) or to low temperatures in areas where seeds could be found (Brisson *et al.* 2008; Bart and Hartman, 2003). Invasive *P. australis* seeds can speed up the germination rate, usually in exposed areas, as they are not good competitors at that stage (Stanton, 2005; Kettenring and Whigham, 2009). For example, in Manitoba, seedlings were observed but failed to survive through to the next growing season (Shay and Shay, 1986). It was also observed that this species is not an important component of the seed bank there. In addition, mature seeds can have very low viability, preventing their germination (Kettenring and Whigham, 2009). Higher temperatures, like those predicted by climate change could mean that more seeds will germinate, hence creating new opportunities for dispersal (Brisson *et al.* 2008). A recent study has shown that seed propagation could be an important vector in long-distance dispersal of the invasive haplotype (Belzile *et al.* 2009). The species, which is already very resistant to control measures, is becoming even more competent at spreading.

Although this species presents broad ecological tolerances and characteristics that favour it in competitive interactions, there are also factors that limit its spread. Water depth is a particularly good indicator of suitable habitat for *P. australis*. This species is restricted when the water level rises for long periods; it will not be able to establish itself in water deeper than 0.5 m unless water level fluctuations occur. Once established in an area, if the water level rises above 1 m, the population will not survive more than three years. In similar conditions, most of the well-established individuals will already be dead after two years (Shay *et al.* 1999). It is intolerant to prolonged flooding (Haslam, 1972; Warren *et al.* 2001) and its cover is reduced when subjected to 100 days or more of flooding per year (Hudon *et al.* 2005). Poor performance of *P. australis* in these limiting conditions is probably due to limited dissolved oxygen in the water when the water level rises (Amsberry *et al.* 2000). Further, it has been suggested that frost slows its progression (Jodoin *et al.* 2008), while severe frost can kill the rhizomes (Haslam, 1972). Cold winter conditions limit the expansion of *P. australis*, particularly at its northern limit in Quebec. Climate warming may alleviate some of these pressures. Lastly, it will not grow where light is limited or where light intensity is low due to shading by other plants (Shay *et al.* 1999). For this reason, forested areas act as a particularly good barrier by limiting light intrusion in the system and providing underground competition for resources (Lelong *et al.* 2009). Of course, *P. australis* is also slowed by some other plant species like *Typha* sp. or *Spartina* sp. due to inter-specific competitive interactions for resources (Amsberry *et al.* 2000).

Competitive Interactions and Plant Invasion

Competition is a fundamental process in plant populations and communities and usually acts among neighbouring individuals for limited resources. Depending on the limitations of the ecosystem services, either above- or below-ground competition is prevalent (Casper and Jackson, 1997). Competition is often thought of as a frequent phenomenon in nature, with one species always displacing another. It can be inter-specific (between the individuals of different species) or intra-specific (between the individuals of same species). Competitive ability can be compared between species in terms of two components: effect and response (Miller and Werner, 1987; Goldberg, 1990). The former refers to the ability of the species to suppress the growth of other individuals through depletion of resources and the latter refers to a species' ability to tolerate resource levels that have been depleted as a result of competition (Goldberg, 1990). The distance between neighbours and their overall spatial configuration are important factors in competition (Stoll and Weiner, 2000).

In most cases, competition implies exploitation of limiting resources that are heterogeneously distributed. Plants compete for limited resources that are essential for their existence – i.e. water, nutrients and light – and actively alter the environmental conditions not only for themselves but also for their neighbours (Stoll and Weiner, 2000). Resource limitation is relatively difficult to study for plants in their natural environments but it is easy for aquatic organisms (e.g. algae) or isolated plant individuals. The fate of a plant depends to a large extent on its local environmental conditions (Aerts, 1999). Some modifications can make it easier for other individuals to establish, survive, grow and reproduce. All these positive effects are referred to as facilitation (e.g. Callaway, 1995; Callaway and Pennings, 2000; Stoll and Weiner, 2000; Callaway et al. 2011a, 2011b).

Competition among plants can, however, be different above and below ground. Both above- and below-ground competition can affect repulsion and aggregation patterns among individuals (e.g. in semi-arid woodland trees) (Martens *et al.* 1997). Plants have adapted to their environment and also compete for light. Several strategies might be successful: overtopping neighbours, using light as effectively as possible, or changing morphology. Traits supporting these strategies include the high plasticity of crown shapes and physiology (Schwinning and Weiner, 1998), the ability of vegetative ramification, the production of more leaf area in the top layers and high turnover rates of leaves (Aerts, 1999).

Some plants may have a high competitive ability in nutrient-rich habitats because they are able to optimise their resource allocation of above-ground biomass and/or increase nutrient uptake from the soil. The role of plant competition in infertile environments is still debated. According to Grime (1988), traits for nutrient retention are far more important than the competitive ability for nutrient uptake in nutrient-poor environments. Tilman (1988) argues that competition is common in fertile as well as in infertile environments, but the relative intensity of above- versus below-ground competition changes along the gradient.

Adaptive root allocation and plasticity in root morphology are thus most important traits for 'searching for nutrients', either by increasing root biomass or by increasing root lengths. These strategies, however, will only be successful in soils that include nutrient-rich spots. Plants species evolve strategies for avoiding competition. Strategies such as long-range dispersal, use of different time niches, or seed production adapted to environmental conditions can decrease resource competition. These different spatial and temporal strategies might not only be the result of competition, but can also be due to other evolutionary processes (Townsend Peterson, 2003).

Two modes of competition are distinguished: symmetric and asymmetric. Symmetric competition is where plants share resources proportionally to their size. With asymmetric competition, larger individuals obtain all the contested resources disproportionally over smaller individuals, leading to a growth depression of the latter (Weiner, 1990; Stoll and Weiner, 2000). It is usually assumed that below-ground competition is size symmetric, whereas light competition is size asymmetric (Schwinning and Weiner, 1998). Mechanisms such as plasticity, allometry and allocation, which might change the degree of asymmetric competition, are still not clearly understood (Schwinning and Weiner, 1998; Bauer *et al.* 2004).

Competitive interactions among neighbouring plants are considered an important determinant of species composition and abundance in natural systems (Reader and Bonser, 1993; Howard and Goldberg, 2001). Alien species have been found to have superior competitive ability, which is considered to be an essential factor for the promotion of invasives (Baker, 1965: Roy, 1990), and competitive exclusion by native plant species seems to be a major force resisting exotic invasions (Keane and Crawley, 2002). Therefore, it might be expected that highly invasive plants should exhibit large competitive superiority and large competitive responses in native vegetation. Additionally, invasive species might be expected to have weaker competitive effects on, and a stronger competitive response to, conspecifics than heterospecifics, resulting in self-facilitation (Wilson and Agnew, 1992; Conway *et al.* 2002). Communities may resist biological invasion through a variety of biotic processes, including competition, herbivory, predation or disease. Still, most ecologists focus on competition as the main source of resistance. Invading species are readily established where levels of competition are low between the pre-existing vegetation and the invader (Crawley, 1986), such as highly disturbed habitats of urban areas that often have much bare ground (Crawley *et al.* 1996).

Many introduced plant species become successful invaders not because they have greater competitive ability than native species, but because they benefit more from disturbed or nutrient-rich conditions associated with human activities (Daehler, 2003). Petryna *et al.* (2002) observed that the establishment and persistence of the plant species *Oenothera erythrosepala* and *Cirsium vulgare* that were introduced from Europe and have become abundant in the mountain grasslands in central Argentina strongly depended on above-ground biomass removal by cutting and burning and soil disturbances. The role of competition in invasive species can be better understood by the study of the ability of the invader to grow and increase in population size in the recipient community, and the recipient community's tolerance of the invading species. For example, establishment of *P. australis* and *Centaurea diffusa,* which are the most problematic introduced plant species in North America, was shown to be enhanced by high resource availability, and both the density and size of these species were increased when competition was also reduced by partially removing resident grass vegetation (Seastedt and Suding, 2007; Smith, 2008).

Introduced species can undergo adaptive evolutionary change, enabling them to establish and spread in their non-native range (Leger and Rice, 2003; Blair and Wolfe, 2004; Bossdorf *et al.* 2005; O'Donnell and Armbruster, 2009; Barney *et al.* 2009; Orians and Ward, 2010) and this influences the outcome of their species interactions. However, few studies have tested whether invasive species can evolve traits that may enhance their species interactions to fuel invasion success (Seifert *et al.* 2009). Invasive species may evolve traits that not only modify their competitiveness, but also generate more interactions with other species. Hawkes *et al.* (2010) found that the degree of enemy release (from insects and pathogens) became more dynamic as the residence time of introduced *Senecio* species in the UK increased. Seifert *et al.* (2009) discovered that introduced North American populations of *Hypericum* have evolved a reduced dependence on their mycorrhizal symbionts.

Most of the studies have focused on the life history characteristics of exotic and native species from different genera and families so as to test whether invasive species outcompete native species and how they do it (Goergen and Daehler, 2001; Katz *et al.* 2001; Blicker *et al.* 2002; Morris *et al.* 2002; Goergen and Daehler, 2002; Daehler, 2003; Gerlach and Rice, 2003; Booth *et al.* 2003). However, such comparisons do not take into account evolutionary background, which may restrict species variability in terms of morphological, physiological or functional traits. For this reason, it may be more meaningful to compare life history traits for phylogenetically related native and invasive taxa under similar environmental conditions so as to find differences in their attributes (Mack, 1996; Radford and Cousens, 2000). As per Vila and Weiner (2004), in order to find out whether alien invasive plants are superior competitors to natives, one should pair native and alien invasive species with the same life form, because a number of studies have reported that species with distinct life forms react very differently to competition (Gerry and Wilson, 1995). Invasives may outcompete native species, replacing or excluding them and, therefore, transforming fundamental ecosystem structures. Invasives may indirectly change the structure and species composition of the ecosystem by altering the way in which nutrients flow through it (McNeeley *et al.* 2001). Serrano-Garcia *et al.* (2007) designed a replacement series experiment in order to establish whether the two invaders were more competitive than their native congener. Results showed that the alien *Senecio inaequidens* was the best performer competitor and *Senecio inaequidens* was the most competitive species, more than the geographically restricted native *S. malacitanus* and the alien *S. pterophorus* both in water-stress and control conditions. Overall, the results showed that biomass production per plant for *S. inaequidens* was greater in mixed rather than in pure cultures, which would indicate that *S. inaequidens* is a successful competitor (Serrano-Garcia *et al.* 2007). Similar results were obtained for *S. madagascariensis* – the invasive diploid taxon of *S. inaequidens* in Australia (Scott *et al.* 1998) – which also grew and reproduced more rapidly than the closely related native congener *S. lautus* (Radford and Cousens, 2000).

Various experiments and field observations have confirmed that the occurrence of alien invasive species can threaten the persistence of native species (Parker and Reichard, 1998), most probably because of the negative impacts of competition. D'Antonio and Mahall (1991) found that the succulent perennial *Carpobrotus edulis* lowers the soil water availability to native shrubs in coastal habitats and this consequently reduced their growth and reproduction. Similarly, Melgoza *et al.* (1990) found that invasion by *Bromus tectorum* reduced the amount of soil water available for other plants in Nevada grassland. In addition, Busch and Smith (1995) presented evidence suggesting that a combination of water and light competition underlie the effects of salt cedar, *Tamarix*, on most of the native riparian trees in the south-western USA. Per-individual or per-amount effects of an alien invasive species on native species, and vice versa, can be estimated with intensity competition indexes (Goldberg and Werner, 1983).

Despite their limitations, competition indexes are powerful tools for quantifying the outcome of competition between native and invasive species (Grace, 1995; Goldberg *et al.* 1999). Competition indexes can quantify the proportional decrease in plant performance due to competition, and compare the effects of competition under different environmental conditions or on different species under similar environmental conditions. Furthermore, competition indexes meet the criteria necessary to compare differences among independent studies (Gurevitch *et al.* 1992; Goldberg *et al.* 1999). A comparative approach has often been useful in helping to understand the mechanism of what determines the patterns of invasion success (Rejmanek, 1995; Mack, 1996; Grotkopp *et al.* 2002). The comparative approach involves pairing native species with invasive species or non-invasive congeners. If a consistent difference can be identified between the native and the invader, this difference might help explain why the invader has become more successful. An invader that usually outperforms co-occurring native species is expected to increase in relative abundance over time, and more abundant invaders are expected to have greater impacts on co-occurring native populations (Daehler and Carino, 1999).

Plant performance can be measured by various traits, including growth rate, competitive ability and fecundity (Goldberg and Landa, 1991). Relative competitive performance was generally based on the biomass accumulation of the invader relative to the native species when grown in competition. As with most other fitness-related traits, the relative competitive performance of invasive versus native species often depended on environmental conditions. Out of

16 studies that experimentally assessed competitive performance, only five (30%) seemed to show globally higher competitive performance by the invader. These five invaders were *Dipsacus sylvestris* (Huenneke and Thomson, 1995), *Bromus inermis* (Nernberg and Dale, 1997), *Spartina alterniflora* (Callaway and Josselyn, 1992) *Pennisetum setaceum* (Carino and Daehler, 2002) and *Lythrum salicaria* (Mal *et al.* 1997).

In several comparisons, competitive performance depended on nutrient availability. Studies on the control of invasive species suggest that competition can have more negative effects on invasive plant growth than herbivory (Muller-Scharer, 1991; Lonsdale and Farrell, 1998; Willis *et al.* 1998). The success of invasive plants depends on the diverse community of pathogens above ground and mutualists that inhabit soils (Klironomos, 2002; Wolfe and Klironomos, 2005; Reinhart and Callaway, 2006; Pringle *et al.* 2009). Exotic species encounter fewer inhibitory effects of soil biota where they are introduced than in their native range (Callaway *et al.* 2004; Hierro *et al.* 2005; Van Grunsven *et al.* 2010). Thus, soil microbes can facilitate invasion in recipient communities while inhibiting plants at home. On the other hand, disturbance, which results in an increase of resources at the individual level and reduced competition, often appears to be necessary for the establishment of many alien species (Hobbs and Huenneke, 1992). Thus, release from competitors is another mechanism that can also contribute to the success of the invasion of a community (Crawley, 1986).

Plant Invasion and Soil Microbes

One of the most exciting aspects of biological invasion studies is how invasive alien plants affect native soil microbial communities in their invaded habitats, as plant–soil microbe interactions play a major role in determining plant community structure (Bever, 2003) and in some cases strengthen the invasive potential of plants through complex feedback loops (Richardson *et al.* 2000). Soil microbial communities may also be affected by the invasion of non-native plant species (Kourtev *et al.* 2002). It has been found that alien plants suppress harmful rhizosphere soil microbiota (Bias *et al.* 2004a; Lorenzo *et al.* 2010) and enrich beneficial bacteria, thereby establishing positive feedback which could contribute to their proliferation (Klironomos, 2002; Batten *et al.* 2006;

Kulmatiski and Beard, 2008; Sanon *et al.* 2012) at the expense of native biodiversity (Callaway *et al.* 2004; Lorenzo *et al.* 2010). Many studies indicate negative effects of invasive plants on soil micro-organisms. For example, a significant reduction in richness of soil fungi due to invasion by *Bromus tectorum* has been reported (Belnap *et al.* 2005); and the dominance of non-mycorrhizal *Alliaria petiolata* (an obnoxious invader of eastern North American hardwood forests) has been shown to cause a rapid decline in the abundance (Roberts and Anderson, 2001) and function (Wolfe and Klironomos, 2005; Stinson *et al.* 2006; Wolfe *et al.* 2008; Callaway *et al.* 2008; Vogelsang and Bever, 2009; Pringle *et al.* 2009) of arbuscular mycorrhizal fungi. Moreover, bacteria as well as soil fungi have been negatively affected due to invasion by *Acacia dealbata* (Lorenzo *et al.* 2010). It has also been reported that certain invasive species secrete allelopathic chemicals that inhibit native mutualisms (Callaway *et al.* 2008; Zhang *et al.* 2010) and some exotic invasive species also accumulate soil pathogens which inhibit native plants (Westover and Bever, 2001; Eppinga *et al.* 2006; Mangla *et al.* 2008; de la Pena *et al.* 2010; Sun and He, 2010; Inderjit and Van der Putten, 2010). For instance, *Cromolaena odorata* accumulates high concentrations of a generalist soil pathogen in its rhizosphere soil, which suppresses native plant species more than the plant itself (Mangla *et al.* 2008).

Soil microbial community is also affected by changes in nutrient pools and other processes, like nitrogen, phosphorus and potassium cycles (Wolfe and Klironomos, 2005; Kao-Kniffin and Balser, 2007; Lorenzo *et al.* 2010). Thus, the changes in root exudates and plant litter are often highly associated with shifts in microbial community composition (Broughton and Gross, 2000; Jones *et al.* 2004; Paterson *et al.* 2007; Nie *et al.* 2010). *Ageratina adenophora* is an aggressive subtropical Mexican invader, which is rapidly spreading in China and India. Through glasshouse studies, Niu *et al.* (2007) found that soil microbial communities associated with native plants exert a positive impact on *A. adenophora* and a negative impact on native plants. Once the invader has established itself in the non-native range, it increases the abundance of fungi relative to bacteria, as well as increasing the abundance of arbuscular mycorrhizal fungi in the soil. The induction of this shift in soil microbial communities by *A. adenophora* can facilitate its growth and suppress native plants.

Morphological Traits Contributing to Invasiveness

Invasive *Phragmites* in India

Native *Phragmites* in Canada

Various morphological traits that influence the invasive attributes of alien plants include growth form, plant height, leaf form and propagule size. A brief overview of the studies that have dealt with these traits is presented here. Since most of the studies pertaining to trait analysis have been carried out on terrestrial species, the inclusion of aquatic invasive species in such analysis could help to fill in the key information gaps. Invasions are driven by multiple factors, and the role of species traits is context dependent and differs with respect to the stage of an invasion and ecological settings (Hamilton *et al.* 2005; Williamson, 2006; Van Kleunen *et al.* 2007; Wilson *et al.* 2007; Pysek and Richardson, 2007; Kuster *et al.* 2008; Pysek *et al.* 2009; Moravcova *et al.* 2010; Hui *et al.* 2011), and the relative performance of native and invasive species can differ with environment (Daehler, 2003). Rapid progress is being made in plant invasion ecology using the traits paradigm (Hamilton *et al.* 2005; Kuster *et al.* 2008).

Research efforts have primarily focused on identifying a syndrome of traits that promote invasiveness of alien species in their non-native range(s). Though a multitude of traits contributing to species invasiveness has been recognised, attempts to identify key traits that promote invasiveness of alien species have not yet ceased (Kolar and Lodge, 2001; Pysek and Richardson, 2007). In fact, a number of studies have been carried out to compare the traits of both invasive alien plant species and non-invasive plant species under common environmental conditions (Pysek and Richardson, 2007). However, progress in the search for traits conferring invasiveness has been slow, probably because most studies of species traits associated with invasiveness have been restricted to relatively simple traits, such as growth form and plant height, that are readily available in floral compendia or other databases for large numbers of plant species. Nevertheless, exceptionally some recent studies (Lake and Leishman, 2004; Hamilton *et al.* 2005; Colautti *et al.* 2006; Mason *et al.* 2008) have taken into consideration a suite of other traits as well. Daehler (2003) reviewed 119 studies, comparing traits between invasive and native plant species, and Pysek and Richardson (2007) reviewed 64 studies comparing life history traits and the invasiveness of alien and non-invasive plant species. These reviews found characteristics like high fecundity that are frequently associated with invasiveness, but they also found some contradictory results. For example, Daehler (2003) did not find any relationship between a fast growth rate and invasiveness, whereas Pysek and Richardson (2007) found that a fast growth rate is associated with invasiveness.

Recently, Van Kleunen *et al.* (2010) performed a meta-analysis of 117 experimental garden or field studies that measured pair-wise trait differences of 196 non-invasive and 125 invasive plant species. They concluded that invasive alien species had higher values for the performance-related traits than non-invasive species and suggested that it might be possible to predict future plant invasions from species traits. Comparisons of native versus invasive species address the question of what traits of invading species increase their fitness over native species, and invasive alien versus non-invasive alien comparisons address the question of what traits have emerged that distinguish successful invaders from alien species that have not invaded successfully (Hamilton *et al.* 2005).

The consistency of differences between invasive and non-invasive species across the different trait categories could reflect the fact that many traits vary in concert, either due to potential trade-offs or because of some composite traits (Grotkopp *et al.* 2002; Westoby *et al.* 2002). For example, leaf photosynthetic capacity per unit leaf mass (trait category ‛physiology’) is frequently positively associated with specific leaf area (trait category ‛leaf-area allocation’; Wright *et al.* 2004; Leishman *et al.* 2007), which is frequently positively associated with high relative growth rates (trait category ‛growth rate’; Lambers and Poorter 1992; Grotkopp *et al.* 2002). High growth rates, in turn, are likely to result in high annual biomass production (trait category ‛size’) and seed production (trait category ‛fitness’). However the findings of meta-analysis demonstrate that invasive species have higher values than non-invasive species for each of the trait categories and that this is in line with recent findings that many invasive plants are, in contrast to the majority of coexisting native plants, at a position along the global multi-trait leaf-economics spectrum that favours fast growth (Leishman *et al.* 2007). In meta-analysis, Van Kleunen *et al.* (2010) revealed clear trait differences between invasive and non-invasive species and that such differences could also partly reflect potential bias towards studying those invasive species as a consequence of their high performance or a bias towards publication of studies that found higher trait values for invasive species rather than for non-invasive species.

FUNCTIONAL TRAITS AND INVASIVENESS

The question of what kinds of functional traits promote plant invasiveness is of utmost significance for an understanding of plant success in general and particularly for understanding the mechanisms of alien plant invasions. For managing alien plant invasions and taking preventive measures against their spread, recognising these plants based on functional traits constitutes an indispensable cornerstone of designing risk assessment protocols (Groves *et al.* 2001). The functional traits of species, especially related to growth rate, size, fitness, physiology and biomass allocation, are most likely to contribute towards invasiveness (Kolar and Lodge, 2002; Grotkopp *et al.* 2002, 2004).

Plants attain greater heights in order to capture the sunlight. Various studies have been carried out regarding the role played by the height of invasive plants and their invasiveness.

Williamson and Fitter (1996) and Crawley *et al.* (1996) found species invasiveness in British alien plants to be associated with a tall stature. It has also been found in certain studies that tallness is associated with species invasiveness (Goodwin *et al.* 1999), either alone or in association with other life history strategy (Pysek and Jarosik, 2005). However, Lloret *et al.* (2005), Hamilton *et al.* (2005) and Cadotte *et al.* (2006) reported no such relationship between height and invasiveness of plants. Alien plants generally do not seem to be taller in their non-native ranges (Thebaud and Simberloff, 2001).

Several studies have reported significant differences between the leaf trait relations of co-occurring native and invasive species. In new environments, exotic plants may be freed from their natural herbivores and thus suffer less herbivory than native species (Strong *et al.* 1984). It is believed that invasive plants may be able to allocate less effort to defence than non-invasive or native species, without risk of loss of biomass to herbivores. As a result, leaves of invasive plants may not have the characteristics such as a tough texture or hairs typically employed to deter herbivores. Thus, it is predicated that invasive species will have soft-textured, glabrous leaves in all disturbance types. Invaders characterised by higher leaf area and greater phenotypic plasticity, appear to be "super invaders" (Daehler, 2003). The superior performance of invasive species is partly attributed to thinner leaves and this therefore minimises the carbon cost per unit of photosynthetic area (Baruch and Goldstein, 1999; Nagel and Griffin, 2001). Most studies have reported that the photosynthetic rates of invasive aliens are higher than those of natives (e.g. Pattison *et al.* 1998; Baruch and Goldstein, 1999; McDowell, 2002), whereas other studies have shown there is no significant difference, or even a higher photosynthetic efficiency in the natives (Smith and Knapp, 2001; Bastlova and Kvet, 2002). Whether the photosynthetic capacity of the invaders is higher than their native congeners seems to be related to the growing conditions (Daehler, 2003; Nagel and Griffin, 2004; Wang *et al.* 2006).

Several studies have shown that invasive species have a greater specific leaf area (SLA) and lower root:shoot ratio and exhibit more rapid photosynthetic machinery compared with native or non-invasive species (Daehler, 2003; Schlaepfer *et al.* 2010; Pysek and Richardson, 2007; Van Kleunen *et al.* 2010). SLA is one of the key traits that regulates and controls plant functions such as carbon allocation and carbon assimilation (Reich *et al.* 1997; Zhang and Feng, 2004; Feng *et al.* 2004). A greater SLA indicates thinner leaves, which may be easier to produce quickly than thicker leaves of a similar surface area (Lambers and Poorter, 1992). A high SLA may therefore be associated with more rapidly growing species that evolve under nitrogen-rich conditions (Lambers and Poorter, 1992; Schippers and Olff, 2000). It has been demonstrated that this trait may be important in determining the success of many exotic species in different tall grass prairie environments. Rejmanek *et al.* (2005) concluded that invasiveness is strongly linked to leaf traits associated with rapid carbon capture (high leaf:area ratio, high SLA and the relative growth rate). However, few studies have tested leaf trait relationships of native and exotic plant species. Pattison *et al.* (1998) compared dark respiration and assimilation rates of four native and five invasive Hawaiian rainforest species and found that the invasives assimilated carbon dioxide at a lower respiratory cost. Gulias *et al.* (2003) found that species endemic to the Balearic Islands had lower assimilation rates for a given SLA or foliar nitrogen content than widespread non-endemic Mediterranean species. Knee and Thomas (2002) found that the net assimilation rate (NAR) is higher in invasive species than in native species. It is another vital trait that is associated with invasiveness of various alien species in physically disturbed habitats and plays an essential role in the establishment of the invasive plants. A clear distribution in the growth form of the species inhabiting older fields (more than 15 years, dominated by perennials having much lower reproductive allocation) and young fields (less than 15 years, dominated by annuals having high levels of relative reproductive allocation) in Minnesota has been found in a study carried out by Gleeson and Tilman (1990).

An increase in the availability of nitrogen and phosphorus as a result of disturbance or reduced uptake by the native species has been broadly identified as a key mechanism facilitating the establishment of annual grasses in nutrient-poor systems (Paschke *et al.* 2000; Beckstead and Augspurger, 2004; Norton *et al.* 2007). A smaller number of structural investments are needed to support growth forms like vines and climbers that allow them to maximise their growth allocation in disturbed sites (Lake and Leishman, 2004). It has also been found that biennial growth form reduces the mortality risks during the various critical life history phases of establishment and germination by seed production and delaying the timing of flowering, and hence becoming short-lived monocarpic perennials (Grime, 1979; Van Bremen, 1984).

Plants produce various propagules for their perpetuation and the continuation of their species but these propagules differ in size. Certain plants produce big seeds but in a small quantity, whereas others produce smaller seeds in a larger quantity. A number of studies have revealed that seeds of a smaller size are positively correlated with invasion success (Hamilton *et al.* 2005) and it has been suggested that large seeds struggle in that respect, because of their physical constraints to emerge from greater burial depth (Thompson and Grime, 1979). Studies have revealed that interaction of the environment with traits of the species influences invasibility (Robinson *et al.* 1995; Burke and Grime, 1996; Higgins and Richardson, 1998; Stohlgren *et al.* 1999; Miller *et al.* 2002). For instance, smaller seed species are more likely to be favoured by disturbance than species with large seeds (Burke and Grime, 1996).

The complex interaction among multiple species traits and the characteristics of the environment mean that invasive alien species do not always show higher performance-related traits than non-invasive or native species (Daehler, 2003). The benefit of having large seeds is that they take advantage in stressful conditions, yield large seedlings, enable better access to light or a more reliable water supply or (both) (Dalling and Hubbell, 2002). There is, however, a trade-off between seed numbers and seed size (Lindsay *et al.* 1999; Dalling and Hubbell, 2002). Thus, the benefits of large seeds arise only after dispersal has occurred. Ambiguous or non-significant results have been obtained by comparing propagule size between alien invasive and several native species (Thompson *et al.* 1995; Williamson and Fitter, 1996). Studies carried out by Cadotte and Lovett-Doust (2001) revealed very small seeds in alien plant species, whereas Crawley *et al.* (1996) reported bigger seeds in alien species, and the probability of aliens having seeds smaller than those of natives was associated with disturbed habitats (Lake and Leishman, 2004). However, it needs to be emphasised that both large and small seeds have their own advantages and disadvantages and their potential benefits to an alien plant are context dependent. For instance, Thompson *et al.* (1993) found that small seeds are easily dispersed by wind and remain viable for longer periods in soil. Henery and Westoby (2001) found that small seeds are correlated with a high annual seed output. Larger seeds, on the other hand, are better for establishment (Harper, 1977) and more attractive to vertebrate dispersals (Richardson *et al.* 2000).

Functional traits provide new insight into how organisms gain, capture and deploy resources within the environmental constraints. Studies have revealed that the suite of traits favouring invasiveness in plant species vary among different habitat types (Pysek *et al.* 1995). Identification of trait attributes of exotic plant species that are correlated with invasiveness is an important step in understanding the dynamics in biological invasions (Baker, 1965; Rajmanek and Richardson, 1996; Alpert *et al.* 2000; Kolar and Lodge, 2001; Richardson and Pysek, 2006; Cadotte *et al.* 2006; Pysek and Richardson, 2007). Research on biological invasions has primarily focused on identifying traits associated with invading species (Kolar and Lodge, 2001; Williamson and Fitter, 1996; Rejmanek and Richardson, 1996) and characteristics related to the invasibility of ecosystems (Crawley, 1987; Lavorel *et al.* 1999). Many findings have reported that trait combinations and interactions can better support invasion success rather than single traits. Comparisons across (1) phylogenetically related native and invasive species and (2) ecologically similar, sympatric non-invasive and invasive species are two important methods used to explore the key traits that are associated with invasiveness (Mack, 1996; McDowell, 2002). Species traits often associated with invasiveness include a short generation time, large seed crops, prolonged seed viability, a long fruiting period, transport by wind or animals, and a broad native range (Pysek *et al.* 1995; Rejmainek and Richardson, 1996; Williamson and Fitter, 1996; Goodwin *et al.* 1999;

Rejmanek, 2000; Kolar and Lodge, 2001). The time that a plant performs various growth and developmental processes, relative to that of its neighbours, determines its success in the invasion process. If a particular invading plant is at the right place, at the right time, it is the traits that it then expresses that make it a loser or winner compared to its neighbour.

DISPERSAL SYNDROMES AND INVASIVENESS

Dispersal is an active process if it involves numerous important behavioural traits (Barton *et al.* 2009) and a passive process if it involves numerous morphological traits, or it can be both. The term "chorology" has been introduced by previous workers for literature use (Van der Pijl, 1972), later replaced by the suffix "chory" to include the agent of dispersal in terms of trait complexity, leading to further exploration of dispersal syndromes. A dispersal syndrome is a complex association of dispersal traits (propagule properties) and reproductive traits (introduction effort) with accurate adaptation to dispersal agent (Cotto *et al.* 2015). Naturally occurring dispersal mechanisms are adapted to enhance seed dispersal (Ridley, 1930; Van der Pijl, 1982). Thus, different structures are easily interpretable in terms of enhancing a certain dispersal mode (dispersal syndromes) (Tiffney, 1984; Hughes *et al.* 1994; Griz and Machado, 2001).

Identifying the dispersal traits in invasion success has remained a challenging task in observational studies because progress towards understanding the mechanism of invasion continuum (Richardson *et al.* 2000) has largely been made by examining invasion correlates, such as residence time (Wilson *et al.* 2007), propagule pressure (Lockwood *et al.* 2005), evolutionary relations (Duncan and Williams, 2002; Proches *et al.* 2008; Diez *et al.* 2008), and geography and species traits (Pysek and Richardson, 2007 Leishman *et al.* 2007; Van Kleunen *et al.* 2010), in isolation rather than together in a single trait investigation.

Despite many traits contributing to the invasion success of alien plants (Pysek and Richardson 2007), an emerging synthesis points to the importance of leaf traits and dispersal syndromes (Rajmanek *et al.* 2005; Richardson and Pysek, 2006). Dispersal syndromes are the sets of traits of fruits or flowers thought to evolve under the selective pressure of seed dispersers ("dispersal syndrome" hypothesis) and pollinators ("pollination syndrome" hypothesis), respectively (Feinsinger *et al.* 1991; Ben-Tal and King, 1997; Forsyth, 2003). Traits related to effective seed dispersal distance are strongly correlated with invasion success (Lloret *et al.* 2005; Hamilton *et al.* 2005; Buckley *et al.* 2006). A small seed size is considered an important trait of successful invaders, through increasing the reproductive output and concomitantly leading to a greater dispersal distance of the species (Rajmanek and Richardson, 1996; Hamilton *et al.* 2005). Often, some seeds are particularly well equipped for dispersal potential by modifying certain structures including wings, floating devices, plumes, hairs, sticky surfaces, hooks and nutritive tissues, which have evolved under selective pressure of the dispersal agent (Traveset *et al.* 2014). Small seeds are positively correlated with increased seed output (Henery and Westoby, 2001), are easily dispersed by wind, and persist for longer periods in soil than large seeds (Thompson *et al.* 1993). Large seeds are better for establishment (Harper, 1977) and are more attractive to vertebrate dispersers (Richardson *et al.* 2000). Studies addressing the effect of efficiency and dispersal mode have not arrived at consistent conclusions (Hamilton *et al.* 2005).

Invasive alien species have been reported to be more likely to be dispersed by humans than native species (Pysek *et al.* 1995; Crawley *et al.* 1996), animals (Cadotte and Lovett-Doust, 2001), wind or water (Crawley *et al.* 1996). In addition to these traits, a growing body of evidence suggests that introduction effort and/or residence time may be of prime importance in biological invasion and it may modify the role of species traits (Wilson *et al.* 2007; Simberloff, 2009). Greater introduction effort is more likely to have a positive effect on the probability that a species will survive environmental and demographic stochasticity (Lockwood *et al.* 2005), and positive relationships between the introduction of alien species and establishment effort have been found both across (Krivanek *et al.* 2006) and within species (Lockwood *et al.* 2005). Additionally, the rate of range expansion (Hulme, 2011) and the probability of regional distribution (Williamson *et al.* 2009) are greater for early introduced species than for those recently introduced. Alien invasive species produce more seeds that are better dispersed (Ordonez and Olff, 2013). This idea confirms that the introduced species spread more rapidly and have an extended range of expansion (Sakai *et al.* 2001; Pysek and Jarosık, 2005). Plant height, seed mass and dispersal syndromes affecting dispersal abilities differ between invasive and non-invasive alien species (Mason *et al.* 2008). Propagule pressure, which is a composite term of propagule size and number, is an introduction effort determining the failure or success of an invasion of alien populations and it is an important trait in determining the rate of geographical spread.

Species with speedy growth and dispersal abilities possess certain traits: a small seed size, a long fruiting period, large propagule production, a short generation time, prolonged seed dormancy, a high rate of biomass production, a short development time to reproductive maturity, dependence on non-specific root symbionts, and they are successful colonisers of disturbed or bare grounds and are pollinators and seed dispensers, demonstrating efficient long-distance dispersal (Alpert *et al.* 2000; Rajmanek, 2000; Williams and West, 2000). Another study suggested that different fruit syndromes might evolve in response to particular dispersal agents (Amico and Aizen, 2000). Other studies have indicated that polychory (dispersal by multiple mechanisms) is common in the dispersal biology of plants (Nathan and Muller-Landau, 2000; Boedeltje *et al.* 2003; Bullock *et al.* 2006). In such cases, seed dispersal occurs by multiple mechanisms, such as a combination of water, wind and animal-mediated dispersal.

Consequently, studies have been using a pool of morphological, phenoplastic, physiological, reproductive and dispersal traits and are trying to prove that differences in such traits contribute to the success of exotic species (Pattison *et al.* 1998; Rejmanek, 2000; Ridenour and Callaway, 2001). As a matter of fact, multiple approaches have been used until now with the objectives of identifying the key traits of successful exotic plant species to determine if these traits differ consistently between invasive and non-invasive alien plant species, which include native-comparison, source-area and target-area methods of analysis (Pysek *et al.* 2004). However, we think it necessary to ensure a clear understanding of these approaches in brief to understand the reason behind applying the target-area approach in our study.

The source-area approach examines the traits of the species passing through early transition phases of transport and establishment in the invasion success (Kolar and Lodge 2001). We find it difficult to elucidate the size of the species and its geographical origin, which are necessary to estimate the chance of dispersal by humans from the source area, otherwise the circumstances associated with dispersal of life history traits associated with the invasiveness of alien species can be confused (Pysek *et al.* 2004).

The native-comparison approach analyses traits of invading species that enhance their potential to increase in abundance over native species in new environments. The approach compares the life history traits of native with non-native species in the new environment. However, among a pool of native species in the study region, some will become invasives as a result of anthropogenic activities, making it difficult to account for any comparison between exotic species and natives (Leishman and Thomson, 2005). Further complexity in applying the approach lies in the evolutionary framework in which the independent evolutionary divergences that can be compared for shifts in life history traits between natives and non-natives is often limited.

In the target-area approach, comparisons are made in the variations between non-native species in a region in terms of their life history traits subject to invasion success (Pysek *et al.* 1995; Rajmanek and Richardson, 1996; Grotkopp *et al.* 2002; Brandle *et al.* 2003; Lloret *et al.* 2004). In the approach it is necessary to work out the traits that accelerate or limit the invasion success. The approach, however, should control the effects of residence time among non-native species to study the comparison in the success or failure based on life history characteristics. This is to reasonably remove the potentially perplexing issue that the longer a species has been in a new environment, the greater the chance that the species has of becoming abundant and widespread (Pysek *et al.* 2003). However, in conformity with the previous studies, we argue that residence time represents the dimension of propagule pressure. If species are present for a long time, more propagules will enter the environment (Rajmanek *et al.* 2005). Consequently, the population will release more propagules into the environment if the species is present in the invaded region for a long time than would be the case if the same species was only present for a short time. We suggest that this could be the reason why residence time had, unlike in previous studies, no effect on the measures of invasion success in recent data, as it was overwhelmed by the effect of population propagule numbers, which facilitated the reproductive traits (Moravcova *et al.* 2015).

Although many different traits may determine the success of exotic species, we examined a subset of dispersal and reproductive traits. Of course, it is worthwhile to mention that functional traits related to physiology, biomass allocation, size, fitness and growth rate promote invasiveness (Kolar and Lodge, 2002; Grotkopp *et al.* 2002, 2004), but the progress in research proposes dispersal characteristics and propagule pressure as key attributes. Therefore, these properties deserve more attention, with the view that they influence many key aspects including evolution of population and diversity, spread of exotic species and dynamics of plant communities (Hovestadt *et al.* 2000b; Hanski, 2001, Simberloff, 2009). In addition, there are several reasons why seed dispersal is important to plants. First, plants must be able to colonise new areas as conditions change because sites suitable for growth vary in time and space (Howe and Smallwood, 1982). Second, dispersal beyond the canopy is a means of avoiding competition between parents and siblings. Finally, the Janzen-Connell hypothesis or "escape" hypothesis (Janzen, 1970; Connell, 1971) indicates that dispersal from the parent is important because seed predators (insects, rodents or microbes) often concentrate their efforts under parent plants where food density is highest. Plants use a variety of strategies for dispersing seeds beyond the range of their branches and while the process is an important part of plant population dynamics, it is especially important if the plants are invasive, exotic species. The comparative approach in invasion studies compares species with common ancestry such as congeners where one species is a successful invader and the other is non-invasive (Mack, 1996). This approach minimises the potentially confounding effects of phylogeny (Clutton-Brock and Harvey, 1991) and clarifies the importance of small differences in the ecology of related taxa in invasion success (Williamson, 1996). The approach could help to test the idea of taxonomic patterns in invasiveness (Rehage, 2003).

The most commonly used classification system of dispersal syndromes is based on the vector of dispersal subject to propagule morphology (Levine *et al.* 2003). The principal agents of dispersal are either abiotic or biotic. Accordingly, the important dispersal syndromes are termed as Anemochory, Hydrochory, Zoochory and Autochory (Levine *et al.* 2003; Van der Pijl, 1982).

Relation of Syndromes with Invasiveness

While identifying the determinants of invasiveness, it has been suggested that single traits are less useful for explaining invasion success than a combination of several traits (Kuster *et al.* 2008). Some dispersal, reproductive and cytological traits were shown to distinguish invasive alien species from non-invasives (Moravcova *et al.* 2010; Kubesova *et al.* 2010). The idea that introduced species have greater seed dispersal distances than natives species underpins much of our understanding of the dynamics of introduced species, including range sizes (Lodge, 1993), rates of spread (Nathan and Muller-Landau, 2000; Pysek *et al.* 2005) and the ability of introduced species to take advantage of colonisation opportunities arising from an increase in resource levels and/or disturbance (Davis *et al.* 2000; Davis, 2003; Didham *et al.* 2005; Colautti *et al.* 2006). The invasiveness of non-native species in new habitats is strongly dependent on the patterns and mechanisms of seed dispersal (Bakker *et al.* 1996; Byers *et al.* 2002a; National Research Council, 2002; Radosevich *et al.* 2003). In order to escape density-dependent mortality, predation and conspecifics competition, and to maintain exponential growth (invasive traits), organisms must be spatially dispersed during their growth, reproduction and death (Holt, 2009; Connell, 1971; Janzen, 1970; Harms *et al.* 2000). Seed dispersal is the means by which plants colonise new habitats (Salisbury, 1961). Density of dispersed seeds increases with proximity to the parent plant. The reduced impact of distance and density-responsive seed and seedling enemies increases propagule survival, resulting in peak recruitment at some distance from the parent and little recruitment near adult conspecifics (Connell, 1971; Janzen, 1970).

Traits related to efficient seed dispersal are also often associated with invasion success (Rajmanek and Richardson, 1996; Hamilton *et al.* 2005; Lloret *et al.* 2005; Buckley *et al.* 2006). The success of an invasion is positively correlated with the number of seeds dispersed into new areas; the possibility that at least some propagules will reach a safe site and lead to successful seed germination increases seed availability (Williamson, 1996; Masters and Sheley, 2001). A small seed size is considered an important trait for a successful invasion, through increasing the dispersal distance of a species and concomitantly a greater fecundity rate (Rajmanek and Richardson, 1996; Hamilton *et al.* 2005). A large number of seeds being produced provides an opportunity for a species to survive demographic and ecological stochasticity (Lockwood *et al.* 2005).

Moreover, multiple introduction points and long-distance dispersal events can increase the rate of invasion and spatially distribute plant populations, such that the extinction of invasive plants at the local level will have a reduced impact at a regional scale (Kiviniemi and Eriksson, 1999; National Research Council, 2002; Fagan *et al.* 2002). Finally, disturbance can increase the availability of safe sites for non-native seeds, in part because native plants show reduced competition (Lepart and Debussche, 1991; Hobbs and Huenneke, 1992; Kotanen, 1997; Seabloom *et al.* 2003). Seed structures and characteristics (mass and shape) affect the efficiency of seed dispersal (Platt and Weis, 1977; Werner and Platt, 1976). Hence, species with different structures form different dispersal syndromes, including zoochory (Andersen, 1993; Beck *et al.* 2010; Bronstein *et al.* 2007; Brown *et al.* 1975; Reichman, 1979- 1981; Samson *et al.* 1992; Sorensen, 1986), anemochory (Augspurger, 1986; Green, 1980; Matlack, 1987), autochory (Ellner and Shmida, 1981), barochory (Leishman *et al.* 1995) and ombrohydrochory (Gutterman, 2002; Van Oudtshoorn and Van Rooyen, 1999). However, so far, very little is known about the evolution of dispersal strategies and dispersal syndromes in plant species of cold deserts. Seed dispersal by frugivores and mammals is an essential process to invade less disturbed habitats (Milton *et al.* 2007). Since long-distance seed dispersal syndromes enhance many key aspects of the biology of plants, including evolution of populations, diversity and dynamics of plant communities, and spread of invasive species (Hovestadt *et al.* 2000a; Hanski, 2001), the study envisages to understand the biology of invasive plants to identify traits contributing to invasiveness, especially in relation to their propagules and dispersal dynamics (Shah *et al.* 2014).

Genetic Basis of Plant Invasiveness

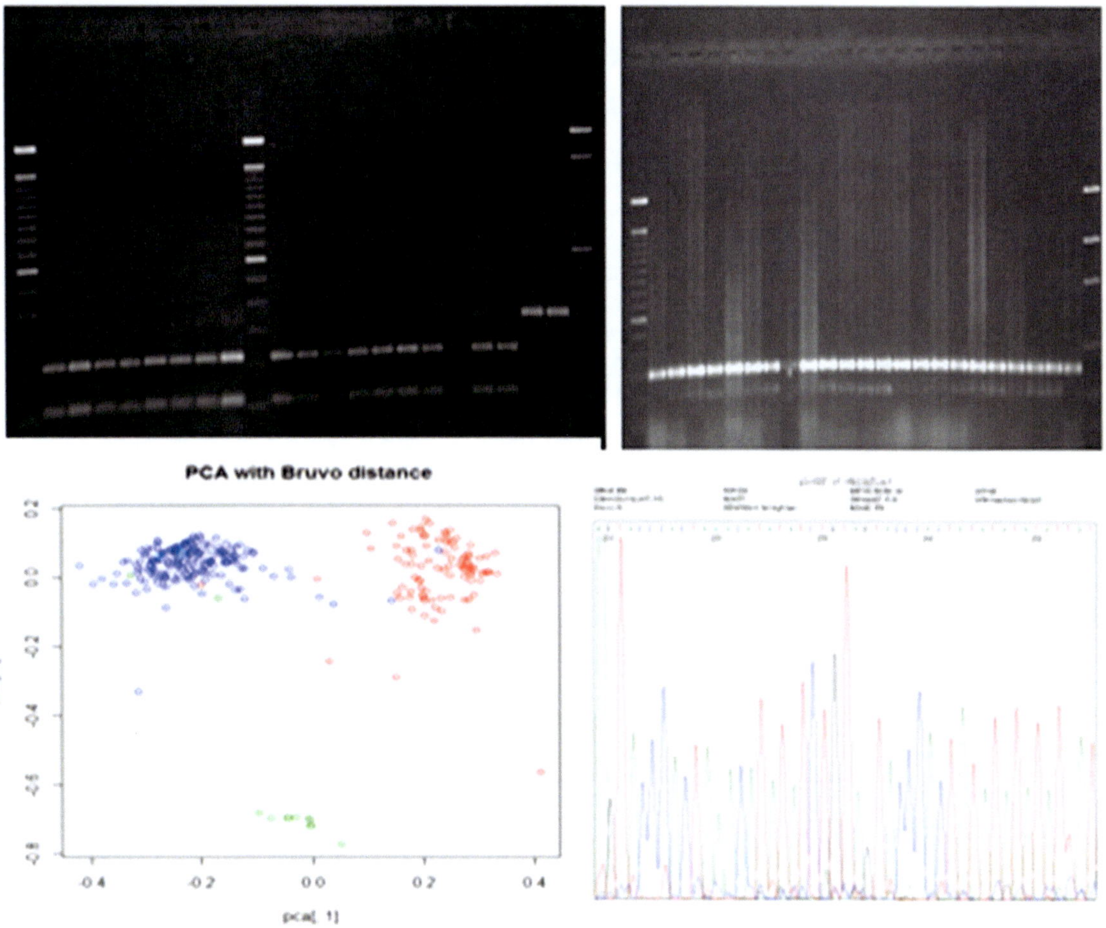

The evolution of morphological, physiological and life history traits and evolutionary dynamics of plant populations have been related to polyploidy (Stebbins, 1950; Thompson and Lumaret, 1992), i.e. the presence of more than two sets of chromosomes either by intra-specific genome duplication (autopolyploidy) or the merging of genomes of different species through hybridisation followed by genome duplication (allopolyploidy). However, only a few studies have assessed the potential significance of polyploidy in the evolution of the ecological and geographical range of the species (Stebbins, 1985). Polyploidy was originally thought to influence the capacity to tolerate environmental stress (Cain, 1944). This was based on the idea that a higher proportion of polyploids, mostly found in herbaceous perennial plants, is generally found at higher altitudes than the related diploids. However, a correlation was found between polyploidy and altitudinal limits in alpine and lowland plants in the Alps (Ehrendorfer, 1980; Thompson and Lumaret, 1992 for reviews). A second hypothesis to explain the widespread distribution of some polyploids, compared to their related diploids, is that polyploids usually colonise new environments exposed by humans or by the retreat of Quaternary glaciations (Hodgson, 1987). Although the comparative studies of geographic distribution among polyploid and diploid taxa in the European floras (Stebbins and Dawe, 1987) have shown that polyploids do often exhibit a wider ecological distribution than diploids, this is not a general trend. Polyploidy also can lead to rapid speciation (Levin, 1983). Unfortunately, no studies have included polyploidy in a comparative study of species richness.

To understand accurately the causal factors affecting species diversity and distribution, it is important to include the ploidy status of the taxa in comparative analysis. However, the difficulty in obtaining information on the ploidy level makes it difficult to do such an analysis across angiosperm families in general. But the possible contribution of polyploidy on the ecological and geographic distribution of species and the species richness of some genera and families in local floras, in comparison with the contribution of other traditionally studied biological characteristics, can be undertaken. According to Levin (2000), the plant species originate, expand, differentiate and, eventually, decline. The newly formed species need to colonise new habitats and establish reproducing populations away from their point of origin in order to succeed. The factors influencing the ability of plants to be efficient colonisers include wide levels of phenotypic plasticity, a high relative growth rate, a high competitive ability, environmental tolerance, an ability to self-reproduce, effective dispersal and avoidance of genetic bottlenecks following founder effects (Levin, 2000). Like all plant species, the invasive species need to be efficient colonisers in new habitats (te Beest *et al.* 2012). However, most of the introduced species fail to establish and spread in new habitats (Williamson and Fitter, 1996; Richardson and Pysek, 2006). Rajmanek and Richardson (1996) highlighted the recent upsurge in research interest on the impact of invasive species on the resident species, communities and ecosystems (Vila *et al.* 2010, 2011).

Polyploidy potentially has important ecological and evolutionary consequences for the fate of introduced plant species (te Beest *et al.* 2012). The newly formed polyploids often have rapid range expansion (Levin, 1983; Hull-Sanders *et al.* 2009; Treier *et al.* 2009) and are over-represented among invasive alien species compared to the native species in some regional floras (Verlaque *et al.* 2002; Pandit *et al.* 2006) and rare plant species (Pandit, 2006; Pandit *et al.* 2011). The empirical data to support this is, however, scarce (Pandit *et al.* 2011). Polyploidy is believed to be the major driver of plant evolution in natural populations (Adams and Wendel, 2005) and it is the single most important mechanism of sympatric speciation in land plants (Otto and Whitton, 2000). Polyploidy also has played a crucial role in the domestication of important crop species, like wheat, maize and cotton (Dubcovsky and Dvorak, 2007). Polyploidisation may alter the plant morphology, phenology, physiology and ecology within a single generation or a few generations, and hence generates individuals that can adapt according to unpredictable environmental conditions, exploiting new niches or outcompeting the progenitor species (Leitch and Bennett, 2007). Taking in view these arguments, it is highly likely that polyploidy may give introduced plant species an important advantage in overcoming novel habitats and assists them in progressing along the introduction–naturalisation–invasion continuum (Richardson *et al.* 2000).

Recent molecular tools have helped in measuring the altered gene expression, regulation and epigenetic structure, resulting in insights into the genetic and genomic consequences of polyploidisation (Paun *et al.* 2010, 2011). Polyploidisation and hybridisation go side by side in the case of allopolyploids, resulting in rapid divergence of the newly formed polyploid and reproductive isolation between the neopolyploid and the parent plants (te Beest *et al.* 2012). Allopolyploidy might act as a genetic stabilising agent to prevent hybrid sterility (Ellstrand and Schierenbeck, 2000), although it appears that most allopolyploids are formed from unreduced gametes and they never go through a diploid phase (te Beest *et al.* 2012). Autopolyploids also experience some genomic changes and this results in increased genetic diversity which is not as drastic as is the case with allopolyploidy (Parisod *et al.* 2010), potentially resulting in altered morphology, physiology and ecology. The growing interest in autopolyploidy, together with advances in the methodology, have led to an increased recognition that autopolyploidy is more common in the naturally occurring populations than was previously thought (Soltis *et al.* 2007). Therefore, autopolyploidy might be an evolutionary dead end (Stebbins, 1950), but, on the other hand, an important driver of genetic diversification and plant evolution (Parisod *et al.* 2010).

The factors linked with invasiveness and species endangerment are of broad interest due to their potential value in explaining the causes and consequences of population status. While previous research has mainly focused on ecological variables, recent work suggests that genetic attributes may be strongly associated with plant species status. Most of the findings demonstrate the importance of genomic attributes as risk factors of vulnerability to invasiveness or endangerment in higher plant species. In the Indian subcontinent and especially in Singapore, a robust relation of ploidy levels with rarity as well as invasiveness has been established (Pandit, 2006). A related study of a single genus (*Clarkia*) confirms the relationship between ploidy levels and range size (Lowry and Lester, 2006), and an even more narrowly focused study by Treier *et al.* (2009) shows how polyploidy has increased within the invasive range of a single species (*Centaurea maculosa*).

There seem to be a number of physiological and ecological adaptations in polyploids (Soltis and Soltis, 2000; Lee, 2002) which, together, may contribute to a species becoming common or invasive. These adaptations can lead to competitive dominance; some studies have shown that polyploids produce a greater amount of biomass in certain habitats, leaving little space for less common diploid species (Pandit *et al.* 2006). Kubatova *et al.* (2008) observed no positive role of polyploidy with the enhanced growth of the invasive *Lythrum salicaria*, a species that also shows a number of ploidy levels in its native range, but only tetraploids were considered to be invasive. Study on *Senecio inaeguidens* has shown no winter survival in the case of diploids, which potentially results in a lower invasive potential than its tetraploids. Among tetraploids of *Senecio inaequidens*, invasive populations have shown higher winter survival rates. This suggests that winter resistance may be an important trait that promotes invasiveness of this species. Study on *Rubus alceifolius* has shown its polyploidy nature, which could help to explain its invasive character in the area of introduction.

Most of the literature on invasiveness implicitly presupposes that polyploidy affects invasiveness; the added vigour conferred by polyploidy can provide an opportunity for invasiveness. For example, it has been reported that noxious weeds are allopolyploid hybrids (Blair and Hufbauer, 2010; Schierenbeck and Ellstrand, 2009) and polyploidy hybrids demonstrated increased heterozygosity and hence were better equipped to deal with inbreeding depression, making them more dominant than non-invasives (Soltis and Soltis, 2000). However, until recently there has been little evidence to justify the assumption that polyploidy affects invasiveness; it is equally plausible that the successive genetic bottlenecks and rapid range expansion experienced in the process of invasion create selective pressures favouring polyploidy. There is increasing evidence that the 'invasive' traits of invasive species may evolve at least in part in response to conditions experienced during invasion (Whitney and Gabler, 2008); whether this holds for ploidy, however, is unclear.

Polyploidy has to be viewed in a wider context along with other factors that are known to influence invasiveness, such as (1) the absence of closely related species in the native flora (Darwin, 1859; Schaefer *et al.* 2011) or taxonomic predisposition (Daehler, 1998; Pysek, 1998; Richardson and Rejmanek, 2004); and (2) possession of certain life history traits like high relative growth rate, effective dispersal strategies, wide ecological tolerance, high reproductive potential, short juvenile stage, high phenotypic plasticity and self compatibility (Baker, 1965; Rajmanek and Richardson, 1996; Pysek and Richardson, 2007). The invasiveness may be indirectly promoted by polyploidy via enhancing the 'evolution of invasiveness'. In their seminal paper, Ellstrand and Schierenbeck (2000) posed the question of whether the invasiveness is born (i.e. released from fitness constraints) or 'made' (i.e. evolved after colonisation). They argued that some invasives are born, as indicated by the published syndrome of life history traits of some species for invasiveness and the pre-adaptation of successful invaders, but they convincingly confirmed that hybridisation also plays a role in invasiveness. Results from studies on polyploid invasive species have shown that both pathways to invasiveness may be important (te Beest *et al.* 2012). Where more than one cytotype occurs naturally in the native range, only polyploids are present in the novel ranges (e.g. Lafuma *et al.* 2003; Mandak *et al.* 2005), suggesting that pre-adaptation of polyploid invaders is important. However, hybridisation of some species (*Fallopia* spp.) in their novel environments (Mandak *et al.* 2003, 2004; Suda *et al.* 2010) suggests that post-colonisation events (evolving invasiveness) might operate as well. This is supported by the fact that both diploids and polyploids experience a long lag phase before they start to spread in the novel ranges (Williamson *et al.* 2009; Gasso *et al.* 2010). This helps plants to cope with the changing environment and maintain their success at the edge of its distributional range (te Beest *et al.* 2012).

Polyploidy, according to Comai (2005), has both costs and benefits. The exact mechanism involved in polyploidisation is poorly understood (Soltis *et al.* 2010), but the fusion of unreduced gametes is paramount (Ramsey and Schemske, 1998). Unreduced gamete formation is under genetic control and heritable, but it may also be affected by environmental stimuli (Parisod *et al.* 2010), like physical stress (e.g. wounding), nutrient stress, habitat disturbance and climatic fluctuations (Veilleeux and Laucer, 1981). Despite a number of potential problems in becoming polyploid, e.g. genomic shock, there is an assumption that polyploidy confers some key advantages, speeding up the evolutionary success, such as gene redundancy that shields polyploids from deleterious recessive mutations and heterosis, or hybrid vigour (te Beest *et al.* 2012).

Mutations producing deleterious recessive alleles have serious negative consequences on the fitness related traits and therefore polyploid lineages should enjoy an immediate benefit by masking these alleles (te Beest *et al.* 2012). This may be crucial in the case of small and isolated populations (Comai, 2005), and it plays a very important role in the initial establishment of founding polyploid populations. The inbreeding depression is initially lower in the case of polyploids than in their diploid progenitors, but deleterious alleles increase in frequency over time and fitness advantages diminish (Otto, 2007). Polyploidy has been shown to play an important role in the reduction of about 50% of inbreeding depression when compared to their diploid progenitors (Husband and Schemske, 1997). The genetic redundancy may in the longer term facilitate adaptive divergence of duplicated genes and hence allow an emergence of evolutionary novelties (te Beest *et al.* 2012).

At the gene level, fixed heterozygosity may follow allopolyploidisation and this has been shown in the case of highly successful allopolyploid species, such as *Tragopogon* (Roose and Gottlieb, 1976). In allotetraploid *T. mirus* and *T. miscellus*, it has been found that 33% and 43% respectively of the loci analysed had different parental alleles. These represent the classic examples of so-called hybrid allopolyploid species with hybrid vigour or heterosis (i.e. an increase in fitness traits such as growth rate, reproductive output and biomass in hybrids, which brings a competitive advantage over diploid parents) (Comai, 2005). Although fixed heterozygosity is limited to allopolyploids, autopolyploid systems are potentially also featured with high heterozygosity (and nearly double the effective population size) than their diploid parents as a result of polysomic inheritance (Parisod *et al.* 2010). The multiple and independent origin of polyploids, which occurs frequently in naturally occurring populations (Soltis and Soltis, 1999), allows for increased genetic diversity as polyploid offspring from different parents may interbreed.

Another important consequence of polyploidisation is the differential gene expression on a per-cell basis, depending on the dosage of genes according to the degree of ploidy (Osburn *et al.* 2003). If the gene expression levels are purely additive, an individual, complete heterozygous should vary strikingly from homozygotes, with the intermediate range corresponding to partial heterozygotes. And, on the other hand, if levels of gene expression are non-additive, then trait values could fall behind or exceed intermediate parental values (Chen, 2007), which renders increased phenotypic plasticity to both autopolyploids and allopolyploids.

The effects of polyploidy on gene expression have been particularly well studied in agricultural crop and model systems, including fruit size in tomatoes and flowering time in *Arabidopsis thaliana* (Osburn *et al.* 2003) and the regulation of developmental processes in maize. For example, it has been shown that about 90% of gene expression in the synthetic allopolyploids of *Arabidopsis* is additive (Wang *et al.* 2006). However, the same study revealed only small effects of genome doubling on gene regulation in autotetraploids (te Beest *et al.* 2012). Other experiments have shown similar patterns (Adams, 2007), suggesting that the effect of genome doubling per se on the gene expression may be limited in the short term. The genetic redundancy in autopolyploids, however, facilitates the adaptive variation (Parisod *et al.* 2010). The relaxed selection will operate on the duplicated copies of genes which leads to increased gene selection and hence phenotypic novelty (te Beest *et al.* 2012). The novel expression patterns of duplicated genes could also result from the tissue specificity of gene expression, which is known as sub-functionalisation, a process believed to have occurred over long evolutionary time frames (Adams *et al.* 2003). The novel and varying phenotypes may predispose the polyploid lineages to successful colonisation in the non-native or novel environments, i.e. pre-adaptation (Buggs and Pannell, 2007). This is supported by the invasion success of the tetraploid spotted Knapweed, *Centaurea stobe* (syn. *Maculata*), in the USA which has been partly attributed to pre-adaption of native European tetraploids to wider climatic ranges, especially drier habitats, compared to diploids. Separate population genetic (Lachmuth *et al.* 2010) and cytogenetic (Lafuma *et al.* 2003) studies have shown that the pre-adapted tetraploids occur in the highland regions in the native range and are more cold-tolerant than the lowland diploid populations (Lafuma *et al.* 2003).

Recent developments in molecular biology have better enabled an exploration of the relationship between the amount of nuclear DNA (genome size), ploidy status, and the invasiveness and other molecular attributes of a species (Ekrt *et al.* 2009). Inherently associated with ploidy is the amount of nuclear DNA or genome size. While the holoploid genome size (or C-value) is used to describe the size of the whole chromosome complement irrespective of the degree of generative ploidy. Genome sizes vary enormously in vascular plants, spanning from 0.075 to 154.23 pg/1C (Bennett and Leitch, 2010), and diploid genomes of one species might even be larger than or equal to polyploid genomes of a different species. DNA thus plays a dual role in heredity: (1) genic, i.e. holding the precise information about the organism's functioning and development, and (2) nucleotypic, i.e. setting thresholds within which the genes can operate and therefore independently constraining the organism's functional traits. The total DNA amount (C-value) rises with increased ploidy, but the increase is often not proportional, resulting in the decline of Cx-values (i.e. genome downsizing; Leitch and Bennett, 2004). The actual effect of genome size on plant traits then depends on the interplay between ploidy and C- and Cx-values. Universal nucleotypic effects among most organisms include a positive correlation between the cell volume and genome size and a negative correlation with the cell division rate (Gregory, 2005).

Genome size-constrained traits related to individual specific leaf area (Morgan and Westoby, 2005; Beaulieu *et al.* 2007a), growth rates, reproductive success and dispersal such as the minimum generation time (Bennett, 1972; Leitch and Bennett, 2007), and seed mass (Grotkopp *et al.* 2004; Beaulieu *et al.* 2007b; Knight and Beaulieu, 2008) may be of particular importance in invasion success. Generally, plants possessing large genomes are constrained by a range of functional traits and the range of ecological options available to them (see the "large M genome constraint" hypothesis; Knight *et al.* 2005). Whereas small-genome herbaceous species can either be ephemeral, annual or perennial, their counterparts with large genomes are slow-growing obligate perennials. Similarly, large-genome species are generally restricted to the production of large seeds with low dispersal abilities and thus have a lower potential to colonise new habitats. Perhaps not surprisingly, endangered (i.e. typically spatially restricted) species tend to have large genomes (Vinogradov, 2003), whereas the opposite is generally true for weedy angiosperms (Bennett *et al.* 1998). Linked to seed traits and growth rates, genome size may also co-determine the ecological strategy that the plant adopts (i.e. competitive, stress-tolerant or ruderal) and the type of habitats where it will thrive (Bennett, 1987).

Attempts to incorporate genome size information into models aimed at predicting the invasive behaviour date back to the late 1990s. Rejmanek (1996) hypothesised that the ultimate determinant of plant invasiveness in disturbed habitats may be a short minimum generation time, which is a genome size-constrained trait. Other ecologically relevant traits associated with small genomes, such as small seeds, high growth rate and high specific leaf area, may further catalyse the spread of invasive species. For example, Grotkopp *et al.* (2002) reported a negative correlation between the invasiveness and genome size in the genus Pinus (particularly in the wind-pollinated species). The same was found in the genus Artemisia, where weediness or invasive behaviour was coupled with lower genome sizes (Garcia *et al.* 2008). Similarly, Chen *et al.* (2010) observed that both C- and Cx-values had significant effects on plant invasiveness in most plant groups, with the exceptions being trees and the family Fabaceae; Gallagher *et al.* (2011) found no relationship between invasiveness and genome size in Australian Acacia species. Chen *et al.* (2010) found that monoploid genome size had a higher predictive value for assessing plant invasiveness than the size of the holoploid genome, and they suggested that the high proportion of polyploids among weedy plants can at least partly be explained by genome downsizing.

Further insight into the role of genome size in plant invasion was gained by Kubesova *et al.* (2010), who compared genome sizes of nearly 100 alien species naturalised in the Czech Republic with their congeners and confamilials not reported to be naturalised or invasive in any part of the world, and concluded that the former had, on average, smaller genomes. They showed that invasive species in the Czech flora did not differ in genome size from naturalised but non-invasive species, indicating that small genomes may predispose alien plants to become established in novel environments, but not necessarily to become invasive (Kubesova *et al.* 2010). Particularly valuable for the elucidation of the effects of DNA mass on invasive potential in plants are species with both non-invasive and invasive genotypes that show intra-specific variation in genome size. Recently, Lavergne *et al.* (2010) showed that invasive genotypes of reed canary grass (*Phalaris arundinacea*) in North America have smaller genomes than their native source populations in Europe. The observed reduction in genome size, which probably occurred over a few generations, was associated with phenotypic changes (e.g. a higher early growth rate) that increased the invasive potential of the species. This study shows that the selective changes in the amount of nuclear DNA can be of paramount ecological and evolutionary importance, leading to genetic novelties and possibly to the establishment of pheno-/genotypes with enhanced invasive ability.

To better understand the rapid evolution and how invasive species adjust to novel environments, population genetic studies should be combined with analysis of spatial phenotypic differentiation. Identifying whether selection or genetic drift is responsible for phenotypic divergence is a complex task, and requires the initial step of comparing spatial genetic differentiation (F_{ST}) with spatial phenotypic differentiation (P_{ST}). Geographic variation in traits with a genetic basis provides compelling evidence for spatial variation in selection regimes

(Haldane 1948; Slatkin 1973). Currently, most accessible tools for examining the contribution of non-neutral evolutionary forces to quantitative trait variation compare the differentiation of quantitative traits (QST or its phenotypic analogue, PST; Spitze, 1993) to the differentiation of putatively neutral molecular markers (FST; Wright, 1951). The adaptive nature of phenotypical differentiation among populations can be tested relative to a neutral model of quantitative genetic divergence, where neutral genetic variation is used as the null expectation for the amount of differentiation among populations in quantitative variance (Spitze, 1993; Whitlock, 1999). Phenotypical differences between populations are often assumed to be adaptive and both molecular and quantitative variations can be influenced by non-adaptive processes involving random genetic drift and gene flow (Wright, 1931; Falconer and Mackay, 1996). Founder effects, bottlenecks and admixture, in particular, can have strong impacts on the distribution of phenotypical variation within and among populations (Lande, 1980; Lynch *et al.* 1999). Under these conditions, it can be especially challenging and interesting to determine whether heritable phenotypical differences between populations represent the outcomes of adaptive or non-adaptive processes.

ROLE OF MICROSATELLITE MARKERS IN PLANT INVASION AND POPULATION GENETICS STUDIES

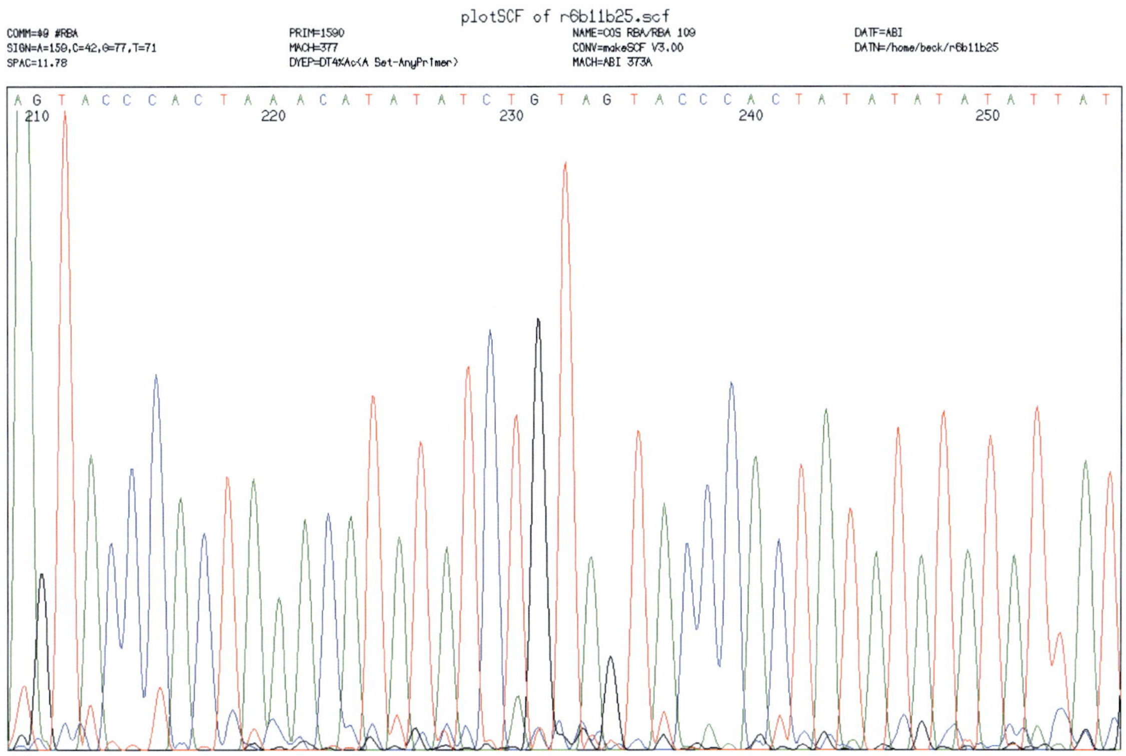

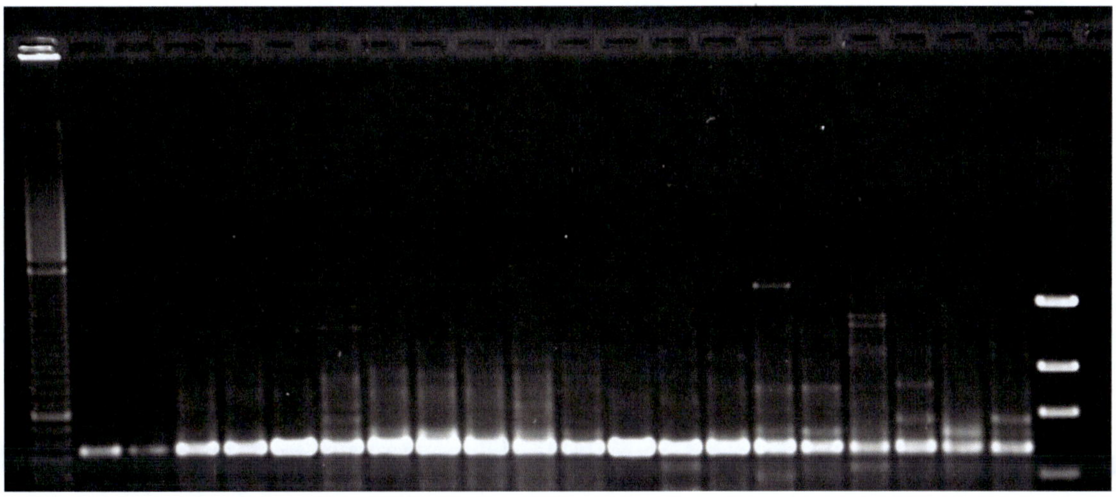

Molecular genetic approaches are powerful tools to investigate the ecological and evolutionary impacts of invasive species. These approaches enable the processes, consequences and future expectations of the biological invasions to be identified. In the past decade, genetic approaches to answering ecological questions have become more efficient, powerful and flexible, and thus more widespread. Genetic analysis also facilitates researchers to reconstruct the histories of the biological invasions (Cristescu *et al.* 2001; Hanfling *et al.* 2002) in that spread patterns and invasion pathways can be identified by comparing the distributions of the particular alleles and the genetic similarity between the native and introduced populations. Population genetic approaches can detect the evidence of population size, migration rates, bottlenecks, kinship, founder effects and more. Several recent reviews detail the myriad of genetic analysis techniques now available and the ecological questions they can address (Bossart and Prowell, 1998; Cruzan, 1998; Davies *et al.* 1999; Sunnucks, 2000; Manel *et al.* 2003, 2005; Beaumont and Rannala, 2004; Pearse and Crandall, 2004). There are several widely used marker types available for molecular ecology studies and many questions can be addressed with more than one type of marker. A comprehensive review of different marker types is provided elsewhere (Avise, 1994; Sunnucks, 2000; Zhang and Hewitt, 2003; Schlotterer, 2004). Microsatellites are of particular interest to ecologists because they are one of the few molecular markers that allow researchers insight into fine-scale ecological questions.

In natural plant populations, microsatellites have great potential for helping us to understand what determines patterns of genetic variation, particularly when used in concert with chloroplast DNA (cpDNA) markers. Their utility has been demonstrated in studies of genetic diversity (Morand *et al.* 2002), mating systems (Durand *et al.* 2000), pollination biology (White *et al.* 2002) and seedling establishment (Dow and Ashley, 1996). Few studies have been carried out using microsatellites in the analysis of the population structure of polyploid species. This is likely due to the problems in analysing polyploid data, as well as difficulties in amplifying loci, possibly because of differences in the parental genomes of polyploids (Röder *et al.* 1995). An area of concern when using microsatellite frequencies to estimate population parameters is size homoplasy, which can cause underestimation of the number of alleles and the levels of heterozygosity and genetic diversity when only allele sizes are considered (reviewed in Estoup *et al.* 2002). A number of studies have demonstrated that microsatellite alleles of the same size can arise from mutation events which either interrupt repeat units or occur in the regions flanking the repeat region. This has been shown to occur both within (Angers and Bernatchez, 1997; Viard *et al.* 1998) and among populations (Estoup *et al.* 1995; Viard *et al.* 1998) and closely related species (Peakall *et al.* 1998; Van Oppen *et al.* 2000). One approach to minimising the risk of misinterpretation of genetic information is to characterise different electromorphs by sequencing, particularly in cases in which other genetic data (e.g. chloroplast or mitochondrial sequences) suggest strong levels of genetic structuring that is not being detected by microsatellite analysis.

Saltonstall (2003a) used 10 microsatellite loci to assess the genetic structure of *Phragmites* populations in North America and Europe at the level of nuclear DNA. Recent evidence based on sequencing of cpDNA indicates that a single non-native lineage of *Phragmites* has been introduced to North America and is likely to be responsible for the dramatic spread of the species across

North America (Saltonstall, 2002). Native individuals persist in many locations, such as in the Midwest and western parts of the country, but introduced populations have overrun the Atlantic coast region and eliminated the majority of native populations (Saltonstall, 2003a). Five diagnostic chloroplast mutations were found to distinguish native and introduced *Phragmites* lineages in North America (Saltonstall, 2002). However, because cpDNA has only maternal inheritance patterns, further analysis of whether the two lineages are hybridising and whether hybrid individuals may be responsible for the widespread invasion and the extent of nuclear genetic variability in the introduced lineage is needed. Srikwan *et al.* (2002) used six microsatellite markers in characterising genetic variability in tree shrews and Tupaia and revealed high allelic variability (A) and heterozygosity (H_o) at these loci. Such characteristics make these genetic markers ideal for linkage mapping and comparative genomics and for studies of pedigree relationships, population structure and population genetic variability in wild populations. Srikwan *et al.* (2002) illustrated another use of these markers in the study of genetic erosion in small, recently fragmented natural populations of tree shrews in south Thailand.

Despite their importance for investigating evolutionary dynamics of bio-invasions, few studies to date have assessed the genetic diversity and structure in the introduced populations and compared them to those in their native range (Giraud *et al.* 2002; Kolbe *et al.* 2004; Maron *et al.* 2004). Similarly, Genton *et al.* (2005b) compared genetic diversity between North American native and French introduced populations of common ragweed to elucidate whether French populations of the Rhone Valley have a single or multiple North American source(s) and where the source population(s) is/are. The results demonstrated that overall genetic variability was similar in North America and in the Rhone-Alpes region, but within the populations, levels of genetic variability were surprisingly lower in the natives than in invasive French populations. Few studies on plant invasions, however, have demonstrated an increase in genetic variability in the introduced range (Squirell *et al.* 2001; Bossdorf *et al.* 2005), while most of the previous plant invasion studies have detected a reduced genetic diversity within introduced populations (e.g. Husband and Barrett, 1991; Amsellem *et al.* 2000) or found no clear differences (e.g. Neuffer and Hurka, 1999).

Erfmeier and Bruelheide (2011) carried out a field sampling campaign on European invasive and native *Rhododendron. ponticum* in Ireland, Georgia and Spain, and North American *R. catawbiense* and *R. maximum* addressed the question of hybridisation on the basis of both molecular marker profiles and quantitative traits of germination success and further ecological and morphological traits. Paul *et al.* (2011) used microsatellite data to compare the genetic diversity of 19 *P. australis* stands in the UK and found that neither the size, management, nor habitat disturbance was affecting genetic diversity. Genetic diversity may therefore be an unpredictable value that is influenced by proximity to other stands; by seed set, which is known to be variable or by availability of optimal seed germination sites such as moist mudflats (Coops and Van der Velde, 1995). The age of a stand can also influence its genetic diversity (Alvarez *et al.* 2005).

Genetic variability is required to permit an adaptive response of the introduced species to the new selective regime imposed by the invaded environment. However, introductions usually consist of a small number of founders, resulting in lower genetic variability in introduced populations compared to native ones (Nei *et al.* 1975). Losses of genetic diversity have indeed been detected both at selective neutral markers (e.g. Amsellem *et al.* 2000; Estoup *et al.* 2001) and at loci under selective pressures (e.g. Eckert *et al.* 1996; Giraud *et al.* 2002). Colonisation following introduction can involve sequential bottlenecks that would further reduce genetic diversity (Amsellem *et al.* 2000; Clegg *et al.* 2002; Estoup and Clegg, 2003; Estoup *et al.* 2004). Loss of genetic diversity during introduction and subsequent range expansion should limit local adaptation possibilities in the new colonised areas (Lambrinos, 2004). Nevertheless, invasive species seem to be able to rapidly adapt to the new conditions they face (Stockwell *et al.* 2003) when introduced (e.g. Lee, 1999; Kinnison *et al.* 2001) as well as when expanding their range, leading to clinal and regional variation in the invaded range.

Several different explanations – depending on the characteristics of the introduced species, the invaded environment and the invasion history – have been invoked to resolve this apparent paradox (Frankham, 2005). First, adaptation may not always be a prerequisite for the successful establishment of introduced populations (e.g. *Caulepra taxifolia* in the Mediterranean Sea; Phillips and Price, 2002). Second, rapid population expansion sometimes occurs just after introduction, retaining genetic diversity (e.g. Zenger *et al.* 2003) and facilitating later adaptation to the environment. Third, polyploidisation and hybridisation phenomena may rapidly produce novel diversity in the introduced area (Ellstrand and Schierenbeck, 2000). Finally, multiple introductions may generate a high level of genetic diversity in the introduced range (Neuffer and Hurka, 1999; Marson *et al.* 2004), sometimes higher than in the invasive range (Kolbe *et al.* 2004).

Regardless of the question, a molecular marker must fundamentally be selectively neutral and follow Mendelian inheritance in order to be used as a tool for detecting demographic patterns, and these traits should always be confirmed for any marker type. Here, we outline the desirable traits of microsatellites compared with other marker types, such as allozymes, amplified fragment length polymorphisms (AFLP), sequenced loci and single nuclear polymorphisms (SNP), focused on both practicalities and ecological considerations.

An ideal marker allows the use of small tissue samples which are easily preserved for future use. In contrast to allozyme methods, DNA-based techniques, such as microsatellites, use Polymerase Chain Reaction to amplify the marker of interest from a minute tissue sample. The stability of DNA compared with enzymes allows the use of simple tissue preservatives (such as 95%-ethanol) for storage. In addition, because microsatellites are usually shorter in length than sequenced loci (100–300 bp versus 500–1500 bp), they can still be amplified with PCR despite some DNA degradation (Taberlet *et al.* 1999). As DNA degrades, it breaks into smaller pieces and the chance of successfully amplifying a segment is proportional to its length. This trait allows microsatellites to be used with fast and cheap DNA extraction methods, with ancient DNA, or with DNA from hair and faecal samples used in non-invasive sampling (Taberlet *et al.* 1999). Furthermore, because microsatellites are species-specific, cross-contamination by non-target organisms is much less of a problem compared with techniques that employ universal primers (i.e. primers that will amplify DNA from any species), such as AFLP. This feature is of particular importance when working with faecal samples or species, such as scleractinian corals, in which endosymbiont contamination is practically unavoidable.

Each marker locus can be considered as a sample of the genome. Because of recombination, selection and genetic drift, different genes and different regions of the genome have slightly different genealogical histories. Relying on a single locus to estimate ecological traits from genetic data creates a high rate of sampling error. Thus, taking multiple samples of the genome by combining the results from many loci provides a more precise and statistically powerful way of comparing populations and individuals. Furthermore, statistical approaches to the questions of most interest to ecologists often require multiple, comparable loci (Pearse and Crandall, 2004). Although Amplified Fragment Length Polymorphism (AFLP), allozymes and random amplified polymorphic DNA (RAPD) techniques are also multi-locus, none of them have the resolution and power of a multi-locus microsatellite study (see Sunnucks, 2000). While AFLP markers can be a good alternative choice to microsatellites (Bensch and Akesson, 2005). Gerber *et al.* (2000) showed that 159 AFLP loci provided slightly less power in determining paternity than six polymorphic microsatellite markers. Sequencing technology has advanced rapidly, but its cost still prohibits the duplication or triplication of workload by using multiple independent gene sequences in parallel (Zhang and Hewitt, 2003).

Microsatellite markers generally have high mutation rates, resulting in high standing allelic diversity. In species for which populations are small or recently bottlenecked, markers with lower mutation rates, such as allozymes, may be largely invariant and only loci with the highest mutation rates are likely to be informative (Hedrick, 1999). A slow mutational process allows the signature of events in the distant past to persist longer. Thus, the selection of loci with high or low allelic diversity will depend on the question of interest. For example, if one is interested in a potential historical barrier to gene flow or tracing the recolonisation of territory since the last ice age, markers with lower mutational rates are likely to be the most informative. In contrast, if one is interested in present-day demography or connectivity patterns, or detecting changes in the recent past (10–100 generations), microsatellites with higher mutational rates are preferable. Questions of paternity or clonal structure are best addressed using microsatellites with the highest allelic diversity, which can provide every individual with a unique genotype identification tag using only a few loci (Queller *et al.* 1993). Similarly, for studies of population structure and migration that employ population allele frequency estimates, the numerous alleles of high diversity microsatellites act as statistical replicates to lend more power to distinguishing populations (Queller *et al.* 1993).

PHRAGMITES AUSTRALIS: A MODEL AQUATIC INVADER

At a large scale, *P. australis* seems to reduce overall biodiversity (Chambers *et al.* 1999; Meyerson *et al.* 2000; Bertness *et al.* 2002). Studies on the impacts of exotic *P. australis* on invertebrates are numerous. Nektonic species do not seem to be affected much by the floral shift as *P. australis* can also be used as a fooder (Meyer *et al.* 2001; Hanson *et al.* 2002; Posey *et al.* 2003; Weis and Weis, 2003). Stem epifaunal communities are less abundant in marshes dominated by *P. australis* compared to those dominated by *Spartina alterniflora* (Weis and Weis, 2003). In addition, *P. australis* affects arthropod communities by removing the herbivores modifying species composition (Gratton and Denno, 2006). Insect species richness and diversity are lower in these communities (Weis and Weis, 2003; Robertson and Weis, 2005), but herbicide spraying seems to restore *Gastropoda* and *Odonata* abundance (Kulesza *et al.* 2008).

A number of studies have shown that *P. australis* seems to have a negative effect on larval and small-fish survival, particularly mummichog larvae (Weinstein and Balletto, 1999; Able and Ragan, 2000; Weis and Weis, 2003; Hunter *et al.* 2006). Other fish species, such as *Morone Americana* and *Anguilla rostrata*, are more abundant in marshes dominated by *P. australis* (Warren *et al.* 2001). One possible explanation for this is that these two species feed similarly in marshes dominated by *Spartina* and *P. australis*, while other fish species do not. As for birds, it seems that *P. australis* is used by some species of birds for nesting, while others are barred from invaded areas (Parsons, 2003). Specialists of short vegetation marshes, such as the saltmarsh sharp-tailed sparrow or seaside sparrow, are negatively affected by the invasion of the common reed. On the other hand, species like the swamp sparrow that like tall vegetation are favoured by reed invasion of marshes. When comparing marshes dominated by *Phragmites* sp. and *Typha* sp., bird diversity is similar. However, a single species, Virginia rail, was strongly linked to the presence of *Typha* (Benoit and Askins, 1999).

Diversity of algae is also greatly affected by the conversion of marshes into being dominated by *P. australis*. *Vaucheria* sp. and *Enteromorpha* sp. are usually less abundant in these marshes, but diversity is restored after herbicide spraying (Warren *et al.* 2001; Kulesza *et al.* 2008). A number of studies of *P. australis* concerning the effect on biodiversity compare communities where *Juncus gerardii*., *Spartina* spp. or *Typha* spp. are the dominant plants, and also where less dominant plant species are affected. It has been observed that species richness is highest in areas where *P. australis* is absent (Warren *et al.* 2001). When present, *P. australis* is associated with other species of low conservation value (Zedler and Kercher, 2004). Because of clonal growth of *P. australis*, available space for co-occurring species is greatly limited; therefore, its stem abundance can be a good marker of wetland diversity and can be used to evaluate the abundance of other species as well. In a European study, this variable was used to explain the variation of abundance (64%) in wetland species (Lenssen *et al.* 2000). When this species is fully dominant in the area, overall site diversity is reduced. When it is partially dominant, diversity is high, possibly due to the addition of opportunistic species to the communities. When absent, the diversity value is medium (Tulbure *et al.* 2007). When comparing the species richness of plant communities, areas invaded by *P. australis* show the lowest richness when compared with the native species (*Impatians*) or with other invasive species (*Lythrum salicaria*).

Human influences, zonation and flood regimes may also have a profound impact on the richness of these communities (McGlynn, 2007). Invasion usually occurs in marshes dominated by short plants like grass species or small sedge and tends to be slowed or stopped by forested areas (Wilcox *et al.* 2003). It has been shown that scrubs and tree seedlings (*A Fraxinus pennsylvanica*,

Baccharis halimifolia, *cer rubrum*, *Liquidambar styraciflua*, *Iva fructescens*, and *Myrica cerifera*) can displace *P. australis* (Havens *et al.* 2003). Thus, *P. australis* could act as an ecological engineer, creating new floating islands and replacing short grass to ultimately be displaced by scrubs and tree species. The capacity of *P. australis* to elevate its environment and dry the ground may provide the stepping stone between low marsh and forested marshes (Esselink *et al.* 2002).

Specific Interactions of *Phragmites* With Other Plants

Impatiens capensis, *Sagittaria latifolia*, *Peltandra virginica* and *Polygonum punctatum* have been found coexisting in plots with *P. australis* and also with *Lythrum salicaria* and *Typha* spp. in brackish environments (McGlynn, 2007). One study reported that *P.australis* is responsible for the displacement of two species: *Solidago sempervirens* and *Atriplex patula* var. *hastata*, also in brackish environments (Minchinton *et al.* 2006). *Panicum virginatum* can coexist with *P. australis* but its cover value does not exceed 20% in brackish water (Warren *et al.* 2001). *P. australis* also displaces *Juncus gerardii* and while coexistence is rare, the latter has been recorded as able to occupy up to approximately 22% of the cover where it occurs in juxtaposition with *P. australis* (Warren *et al.* 2001; Silliman and Bertness, 2004; Minchinton *et al.* 2006). In some cases, surveys indicate that *Juncus gerardii* is able to slow the progression of *P. australis* (Amsberry *et al.* 2000; Minchinton and Bertness, 2003).

In freshwater, *Bidens cernua* was completely excluded by *P. australis*, while the abundances of *Impatiens capensis*, *Schoenoplectus tabarnaemontani* and *Sagittaria latifolia* were considerably reduced. Denis (2011) reported that species richness and evenness remained similar between both samplings of the study (2001–2004) and the Jaccard similarity index between the communities was 18.9%, suggesting a major community structural shift due to the progression of both *Typha* sp. and *P. australis* (Tulbure *et al.* 2007). *Myrica gale* and *Carex aquatilis* are usually less abundant in freshwater plots dominated by *P. australis* (Richburg *et al.* 2001). Communities dominated by *Lythrum salicaria* have a greater species richness. In addition, *Lythrum salicaria* density was found to be similar in communities it dominates and communities dominated by *P. australis* (Keller, 2000). However, these were static observations and do not provide insights on community alteration mechanisms. Reed is slowed by *Spartina patens* and *S. alterniflora* in low marsh habitats (progression of only 6 cm per year in comparison to 25 cm in high marsh habitats) undergoing daily flooding, creating a lack of dissolved oxygen (Windham and Lathrop, 1999; Amsberry *et al.* 2000; Minchinton and Bertness, 2003; Rickey and Anderson, 2004).

However, coexistence with *P. australis* was reported to be limiting *S. alterniflora* cover at 20% (Warren *et al.* 2001). Moreover, there is cartographic evidence of slow displacement of *Spartina* sp. (Havens *et al.* 2003). In nitrogen-rich conditions (14 mg/L), *P. australis* performs better than *S. alterniflora* (Ravit *et al.* 2007). Because the requirement for nitrogen is lower for *Spartina* species, the increase in availability favours *P. australis* and promotes its rapid growth (Meyerson *et al.* 2000). Drained conditions with high levels of available nitrogen will further favour rhizomal growth of *P. australis* (Ravit *et al.* 2007). Rhizotoxicity caused by mesoxalic acid (UV-degraded gallic acid produced by *P. australis*) could also partly explain the competitive advantage of *P. australis* (Rudrappa *et al.* 2009). Both *S. alterniflora* and *P. australis* are used as food by local fauna (Weis and Weis, 2003). In this case, the levels of salinity and water would most probably be the limiting factors for *P. australis*.

Typha species are usually linked with communities that have greater plant diversity in comparison to communities with *P. australis* (Keller, 2000; McGlynn, 2007). Several functions in the ecosystem services are comparatively similar between the two community types. Macro-invertebrate fauna and nesting birds are comparable (Benoit and Askins, 1999; Kulesza *et al.* 2008). *Typha* species prefers silty soil to organic soil, which is the opposite of *P. australis* soil requirements (Tulbure *et al.* 2007). When comparing both species in a constant deluge condition, *P. australis* shows 56% less biomass than *Typha* (Boers and Zedler, 2008). Succession of *P. australis* in *Typha* spp. communities has been observed (Arzandeh and Wang, 2003). In fact, *P. australis* shows a highly significant spatial distribution pattern, particularly in contact with *Typha* spp. The number of shoots ranges from none to an almost saturated value at the zone of contact (Keller, 2000). In low marshes, where both species are present, the stem density of *P. australis* is lower than that of *Typha* (Amsberry *et al.* 2000). In drainage ditches, *P. australis* frequently creates denser colonies than *Typha* species. Where there is competition for space, both species shows reduced biomass but *P. australis* shows a greater level of phenotypic plasticity, enabling it to invade different environments more easily than *Typha* spp. (Bellavance, 2007). While salinity (NaCl) levels of 5 g/L allow it to perform in a superior way to *Typha* spp. Success rates of *P. australis* invasions in *Typha* communities are comparatively higher in stable water levels (Coops and Van der Velde, 1995). This clearly indicates that stress tolerance and water level are of prime importance in the invasion success or failure of *P. australis*.

Saltonstall (2002) identified 27 haplotypes of *P. australis* worldwide, based on the two cpDNA sequences. There were 13 haplotypes (A–I, AA, M, S and Z) found in North America, 10 (I, J, L–Q, U, W and X) in Australia and Asia, five (L–O and T) in Europe, four (K, M, R and V) in Africa, and two (I and Y) in South America (Saltonstall, 2002). The high frequency of haplotype M in Europe and Asia suggested that the rapid spread of *P. australis* in North America from the mid-20th century onwards was caused by the introduction of this Eurasian haplotype (Saltonstall, 2002, 2003b). Similar results were confirmed in Quebec, Canada (Lelong *et al.* 2007), where the invading M haplotype replaced the local haplotypes, undergoing rapid population and range expansions. Two haplotypes of *P. australis* have been reported in Quebec: a native haplotype (also known as *P. australis subsp. americanus*) and an exotic haplotype (also known as *Phragmites australis subsp. australis*). Genetic studies have revealed that many haplotypes of this species can be found in North America (Saltonstall, 2002, 2003a, 2003b; Howard *et al.* 2008). However, it would appear that there is only one native haplotype, F, and one exotic haplotype, M, in Quebec (Saltonstall, 2002). Although *P. australis* subsp. *americanus* (native) and *P. australis* subsp. *australis* (exotic) are the same species, they have been documented as different subspecies (Saltonstall *et al.* 2004) and some traits are different; mainly the lengths of the glumes, ligule and lemnas are significantly longer for the exotic haplotype. In Quebec, the native haplotype is characterised by a red stem and ligule base with fewer hairs on the ligule compared to the exotic haplotype with its green stem and greater number of hairs on the ligule. Moreover, the inflorescence of the exotic haplotype is denser relative to that of the native haplotype, containing fewer seeds (Blossey, 1999 Saltonstall *et al.* 2004).

Although their environmental requirements are usually similar, the exotic haplotype outperforms the native haplotype under high salinity. In a greenhouse experiment, the exotic haplotype survived in concentrations as high as 0.41M NaCl, while the native haplotype did not survive above 0.13M NaCl (Vasquez *et al.* 2005). The native haplotype seems to prefer undisturbed habitats, while the exotic haplotype is frequently associated with roadsides and other disturbed habitats (Lambert and Casagrande, 2006; Jodoin *et al.* 2008; Meadows and Saltonstall, 2009). Water level is another important factor, since the native haplotype shows greater tolerance to high water levels compared to the exotic haplotype (Yulan *et al.* 2008; Meadows and Saltonstall, 2009). It is thought that the growth dynamics of the two haplotypes differ significantly (Hanson *et al.* 2002). It has been shown that colonies of the exotic haplotype have higher above- and below-ground biomass (Hanson *et al.* 2002), especially under nutrient-rich conditions (Saltonstall and Stevenson, 2007). The exotic haplotype also shows greater leaf fresh weight, leaf cover, shoot fresh weight and shoot height (League *et al.* 2006; Meadows and Saltonstall, 2009). Furthermore, the exotic haplotype produces a larger total specific leaf area and has more efficient photosynthesising machinery, partly because it contains twice the amount of chlorophyll content. Due to high levels of available light, photosynthesis is about 50% greater in the native haplotype than the exotic haplotype. This also implies that the native haplotype has a higher demand for total nitrogen and lower water-use efficiency in order to produce its maximal canopy cover. When compared to the exotic haplotype, the native haplotype appears most efficient in terms of resource use (Packett and Chambers, 2006; Mozdzer and Zieman, 2010). Nitrogen availability could be a major factor in defining the most efficient haplotype in spatial interactions.

According to the herbarium record from 1910 to 1960, the distribution of the native haplotype seems to be stable, probably because its ecological niche has been filled. This is not the case for the exotic haplotype, which is in continual expansion (Lavoie *et al.* 2003; Jodoin *et al.* 2008; Meadows and Saltonstall, 2009). Some of the direct impacts of the invasive *P. australis* haplotype on its native counterpart include the following:

1) The invasive haplotype can eliminate the native haplotype by outcompeting it. Competition for similar ecological attributes will drive the local extinction of the native reed. Data from herbarium specimens and surveys has revealed that the native reed may be abundant locally, but it is rare on a large scale (Saltonstall, 2002; Meyerson and Mooney, 2007; Meyerson *et al.* 2000. A study that investigates the interaction between both the haplotypes is needed in order to assess whether the native haplotype is displaced by the exotic haplotype at the population level.

2) Invasive species do not always outcompete native species for resources and dominate space; where its populations are stable and well established and where there are no disturbances at all, natives can outcompete invasives. However, due to its plasticity and the aggressive character of the invasive type, this occurs very rarely

3) Coexistence of both subspecies, but at reduced density for both can be possible where they co-occur.

The invasion biology of both types of haplotype provides a very interesting story for understanding the dynamics that trigger the impacts of invasive species on plant communities. A fine-scale study could prove to be an efficient way to understand the dynamics of invasiveness and the effects caused by the replacement of a native haplotype by an exotic haplotype.

CONTROL AND MANAGEMENT OF AQUATIC INVASIVE SPECIES

Preventing introductions of potentially harmful species is the most efficient way to reduce the threat of Aquatic Nuisance Species., and is a primary focus of the Alian Invasive Species (AIS) Program. Identifying high-risk pathways of introduction and actions to interdict those pathways is integral to prevention. Hazard Analysis and Critical Control Point (HACCP) planning is a management tool that provides a structured method to identify risks and focus procedures on natural resource pathway activities.

After direct prevention measures, the most cost-effective way of stopping a new introduction from becoming an invasion is rapid response, which is dependent on early detection.

Early detection/monitoring capabilities allow us to detect new invasions before they can become established, increase the feasibility of eradication by catching infestations early, and ultimately protect our conservation investments and save long-term control and management costs. Developing partnerships and providing tools to enhance capacity to find, report and ultimately eradicate new invaders, particularly in light of climate change impacts, are critically important to minimising the ecological and economic impacts of invasive species.

Monitoring and reporting locations are also vital for preventing the spread of aquatic invaders. When high-risk or well-travelled paths are identified during monitoring activities, outreach efforts can be targeted at these pathways.

Early detection and rapid response plans also play an important role in prevention. There are several model early detection and rapid response plans available for states or other groups to modify for their specific needs. Many states already have generic and species-specific rapid response plans in place or under development.

When an aquatic invasive species (AIS) has become established in a new location and eradication is no longer feasible, preventing its spread and lessening its impact through control and management becomes the focus. Control and management mean much more than just physically harvesting or otherwise trying to destroy the species. Comprehensive control and management of AIS includes diverse objectives such as eradication within a localised area, limiting dispersal, reducing impacts, education and outreach, and even research on control methods.

Because invasive species span geographic and jurisdictional boundaries, their control and management require coordinated action. Information concerning the distribution, abundance, rates of spread and impacts of the species is critical to regional planning and management efforts. Understanding the diverse ecological, economic and social impacts of AIS is important to prioritising control and management options.

Varieties of control "tools" are needed to contain remove and assess AIS populations and guide management decisions. These tools are applied within an overall integrated AIS management strategy that can be adjusted through an adaptive management process. Actual control of AIS can include physical restraints, removal, pesticides, biological controls and reproduction interference. There are many species that are not currently being controlled or managed in any way due to their widespread occurrence, or lack of effective control and management techniques.

A key to addressing the problems caused by AIS is increasing public awareness of AIS, their impacts, and what individuals can do to prevent their introduction and spread. However, reaching each person whose activities may affect our natural environment is a daunting task. Collaboration, cooperation and coordination across federal and state agencies, local governments, tribal entities, and the public and private sectors is required and facilitating this effort is a primary goal of the aquatic invasive species (AIS) Program. To accomplish this, our Regional AIS Coordinators have:

- developed informational websites;
- conducted workshops; and,
- created outreach materials for national distribution, including:
    - travelling displays;
    - exhibits;
    - pamphlets;
    - AIS identification cards;
    - fact sheets, some in several languages; and,
    - videos.

CONCLUSION AND FUTURE PERSPECTIVES

Assessing the potential for species invasion into new areas gives us tools to recognise potential invaders and favourable habitats for invasion. While humans have greatly facilitated the invasion of aquatic alien species through internationalshipping and trade, aquarium releases and canal construction, etc. (Rahel, 2002), flowing river waters act as natural conduits for the transport of propagules of alien species (Lavoie *et al.* 2003) across geographical and political boundaries.

The important characteristics associated with invasiveness are small, short-lived seeds, a high dispersal rate, high fecundity, a fast growth rate, early fruiting, vegetative reproduction, a wide range of tolerance to environmental conditions, a short generation time, large seed crops, prolonged seed viability, a long fruiting period, transport by wind or animals and a broad native range. The success of invasion relies on many other factors, like the broad tolerance to environmental stresses, release from enemies, hybridisation, efficient use of nutrients and allelopathy. These are some of the possible explanations that provide insights on the success of an invasive species (Galatowitsch *et al.* 1999; Zedler and Kercher, 2004).

In the Indian lakes and wetlands, a reduced quantity of herbivores, increased nutrient inputs and vegetation removal were all considered to be factors enhancing invasive capacities of *Lemna sp*, *Azolla sp*, *Typha X glauca*, *P. australis* and *Myriophyllum spicatum.* Wetlands with hydrological disturbances are more susceptible to invasion and are characterised by a downturn in diversity of native species (Zedler and Kercher, 2004). Habitat alteration facilitates invasions and exacerbates their impacts by increasing the connectedness of watersheds, thereby reducing native competitors and predators and increasing habitat homogenisation to the benefit of opportunistic invaders (Scott, 2006; Johnson *et al.* 2008).

In general, invasive species exhibit a preponderance of performance-related traits compared to the native and non-invasive species. Determining the factors associated with invasiveness will help us to predict future plant invasions from species traits. The various morphological traits influencing the invasive attributes of alien plants include: growth form, plant height, leaf form, propagule size and high fecundity. The functional traits of species, especially those related to physiology, biomass allocation, growth rate, size and fitness, are most likely to promote invasiveness (Kolar and Lodge, 2002; Grotkopp *et al.* 2004). Functional traits provide new insight into how organisms gain, capture and deploy resources within environmental constraints. Studies have revealed that a suite of traits favouring invasiveness in plant species vary among different habitat types (Pysek *et al.* 1995). Comparisons of native versus invasive species address the question of what traits of invading species increase their fitness to thrive over native species, and invasive alien versus non-invasive alien comparisons address the question of what traits have emerged that distinguish successful invaders from those that are unsuccessful (Hamilton *et al.* 2005).

Identifying dispersal traits in invasion success has remained a challenging task in observational studies because progress towards understanding the mechanism of invasion continuum has largely been made by examining invasion correlates, such as residence time, propagule pressure, evolutionary relations geography and species traits, in isolation rather than together in a single trait investigation. A small seed size is considered an important trait of successful invaders, through increasing the reproductive output and concomitantly leading to a greater dispersal distance of the species. Often, some seeds are particularly well equipped for dispersal potential by modifying certain structures including wings, floating devices, plumes, hairs, sticky surfaces, hooks and nutritive tissues, which have evolved under selective pressure of the dispersal agent (Traveset *et al.* 2014). Small seeds are positively correlated with increased seed output, are easily dispersed by wind, and persist for longer periods in soil than large seeds. Species with speedy growth and dispersal abilities possess certain traits: a small seed size, a long fruiting period, large propagule production, a short generation time, prolonged seed dormancy, a high rate of biomass production, a short development time to reproductive maturity, dependence on non-specific root symbionts, they are successful colonisers of disturbed or bare grounds and are pollinators and seed dispersers, demonstrating efficient long-distance dispersal.

Besides conventional morphological, reproductive and physiological traits, recent developments in molecular biology have enabled an exploration of the relationship between the amount of nuclear DNA (genome size), ploidy status and the invasiveness of a species. The newly formed polyploids often have a rapid range expansion and are over-represented among invasive alien species compared to the native species in some regional floras. Polyploidisation may alter the plant morphology, phenology, physiology and ecology within a single generation and hence generates individuals that can adapt according to unpredictable environmental conditions, exploiting new niches or outcompeting the progenitor species. Taking in view these arguments, it is highly likely that polyploidy may give introduced plant species an important advantage in overcoming novel habitats and assists them in progressing along the introduction–naturalisation–invasion continuum.

In natural plant populations, microsatellites have great potential for helping to understand what determines patterns of genetic variation, particularly when used in concert with chloroplast DNA (cpDNA) markers. Their utility has been demonstrated in studies of genetic diversity, mating systems, pollination biology and seedling establishment. The lowest levels of genetic diversity were observed in invasive populations, while the highest levels were recorded in native populations. Using the microsatellite markers for population genetics and phylogeographic studies on invasive species will help us to understand the genetic basis of differential behaviours of invasive species along different stages in non-native regions and help to combat the threat of invasive species that seriously confronts India. Comparing the ploidy status and genetic diversity of invasive and non-invasive plants in native and non-native regions can yield useful insights, with potential implications for biodiversity managers to identify alien invasive species that are likely to become invaders in the near future, thereby prioritising them accordingly for different management strategies.

ACKNOWLEDGMENTS

Financial support by DST-SERB, Govt. of India under the National Post-Doctoral Fellowship (PDF/2018/001689) to GAW is duly acknowledged. The Financial support by the DBT through its Indo-Canadian project (DBT-IC-IMPACT/MAS/30/2015-16) to MAS is also acknowledged. Thanks are due to Mr. James Baron (England) for help in proofreading and language editing.

BIBLIOGRAPHY

Able KW and Ragan SM. (2000). Effects of common reed (*Phragmites australis*) invasion on marsh surface macrofauna: Response of fishes and decapod crustaceans. Estuaries, 23:633-646.

Adams JB and Bate GC. (1999). Growth and photosynthetic performance of *Phragmites australis* in estuarine waters: a field and experimental evaluation. Aquatic Botany, 64**:** 359-367.

Adams KL and Wendel JF. (2005). Polyploidy and genome evolution in plants. Current Opinion in Plant Biology, 8: 135–141.

Adams KL, Cronn R, Percifield R and Wendel JF. (2003). Genes duplicated by polyploidy show unequal contributions to the transcriptome and organspecific reciprocal silencing. Proceedings of the National Academy of Sciences, USA, 100: 4649–4654.

Adams KL. (2007). Evolution of duplicate gene expression in polyploid and hybrid plants. Journal of Heredity, 98: 136–141.

Aerts R. (1999). Interspecific competition in natural plant communities: mechanisms, trade-offs and plant–soil feedbacks. Journal of Experimental Biology, 50: 29–37.

Alpert P, Bone E and Holzapfel C. (2000). Invasiveness, invisibility and the role of environmental stress in the spread of non-native plants. Perspectives in Plant Ecology, Evolution and Systematics, 3: 52–66.

Alvarez MG, Tron F and Mauchamp A. (2005). Sexual versus asexual colonization by *Phragmites australis*: 25-year reed dynamics in a mediterranean marsh, Southern France. Wetlands, 25: 639-647.

Alvarez-Cobelas M, Sánchez-Carrillo S, Cirujano S and Angeler D. (2008). Long-term changes in spatial patterns of emergent vegetation in a Mediterranean floodplain: natural versus anthropogenic constraints. Plant Ecology, 194: 257-271. **Amico G and Aizen MA.** (2000). Mistletoe seed dispersal by a marsupial. Nature, 408: 929-930.

Amsberry L, Baker MA, Ewanchuk PJ and Bertness MD. (2000). Clonal integration and the expansion of *Phragmites australis*. Ecological Applications, 10:1110-1118.

Amsellem L, Noyer JL, Bourgeois TL and Hossaert-Mckey M. (2000). Comparison of genetic diversity of the invasive weed *Rubus alceifolius Poir*. (Rosaceae) in its native range and in areas of introduction, using amplified fragment length polymorphism (AFLP) markers. Molecular Ecology, 9: 443–455.

Andersen UV. (1993**).** Dispersal strategies of Danish seashore plants. Ecography. 16(4): 289–298.

Angers B and Bernatchez L. (1997). Complex evolution of a salmonid microsatellite locus and its consequences in inferring allelic divergence from size information.Molecular Biology and Evolution, 14: 230–238.

Arzandeh S and Wang JF. (2003). Monitoring the change of *Phragmites* distribution using satellite data. Canadian Journal of Remote Sensing, 29:24-35. **Augspurger CK. (1986).** Morphology and dispersal potential of wind-dispersed diaspores of eotropical trees. American Journal of Botany, 73(3):353–363.

Avise JC. (1994). Molecular Markers, Natural History and Evolution.Chapman and Hall, New York, USA.

Bais HP, Vepachedu R, Gilroy S, Callaway R and Vivanco JM. (2003). Allelopathy and exotic plant invasion: From molecules and genes to species interactions. Science, 30:1377-1380.

Bais HP, Park S, Weir TL, Callaway RM and Vivanco JM. (2004a). How plants communicate using the underground information super highway. Trends in Plant Science, 9: 26-32.

Baker HG. (1965). Characteristics and modes of origins of weeds. In the genetics of colonizing species (eds.H.G. Baker and G.L. Stebbins). Academic Press London, 141-172.**Bakker JP, Bakker ES, Rosen E, Verweij GL and Bekker RM.** (1996). Soil seed bank along a gradient from dry alvar grassland to Juniperus shrubland. Journal of Vegetation Science, 7:165–176.

Barney JN, Whitlow TH, Di Tommaso A. (2009). Evolution of an invasive phenotype: shift to belowground dominance and enhanced competitive ability in the introduced range. Plant Ecology, 202: 275– 284.

Bart D and Hartman JN. (2003). The role of large rhizome dispersal and low salinity windows in the establishment of common reed, *Phragmites australis,* in salt marshes: new links to human activities, Eustaries, 26:436-443.69.

Barton KA, Phillips BL, Morales JM and Travis JMJ. (2009). The evolution of an 'intelligent' dispersal strategy: biased, correlated random walks on patchy landscapes. Oikos, 118:309–319.

Batten KM, Scow KM, Davies KF and Harrison SP. (2006). Two invasive plants alter soil microbial community composition in serpentine Grasslands. Biological invasion, 8: 217-230.

Bauer S, Wyszomirski T, Berger U, Hildenbrandt H and Grimm V. (2004). Asymmetric competition as natural outcome of neighbor interactions among plants: results from the field of neighbourhood modelling approach. Plant Ecology, 170: 135–145.

Baruch Z and Goldstein G. (1999). Leaf construction cost, nutrient concentration, and net CO2 assimilation of native and invasive species in Hawaii.Oecologia, 121:183–192.

Bastlova D and Kvet J. (2002). Differences in dry mass partitioning and flowering phenology between native and non-native plants of purple loosestrife (*Lythrum salicaria* L.). Flora, 197: 332–340.

Beaulieu JM, Leitch IJ and Knight CA. (2007a). Genome size evolution in relation to leaf strategy and metabolic rates revisited. Annals of Botany, 99:495–505.

Beaulieu JM, Moles AT, Leitch IJ, Bennett MD, Dickie JB and Knight CA. (2007b). Correlated evolution of genome size and seed mass. New Phytologist, 173: 422–437.

Beaumont MA and Rannala B. (2004). The Bayesian revolution in genetics. Nature Reviews. Genetics, 5: 251–261.**Beck M. J, Vander Wall S. B.** (2010). Seed dispersal by scatter-hoarding rodents in arid environments. Journal of Ecology. 98 (6): 1300–1309.

Beckstead J and Augspurger CK. (2004). An experimental test of resistance to cheatgrass invasion: limiting resources at different life stages. Biological Invasions, 6: 417–432.

Bellavance ME. (2007). Competition interspecifique et plasticite morphologique chez *Phragmites australis* et le complexe *Typha latifolia – Typha angustifloia.* In Science Biologique, (Montral University de Montreal).

Belnap J, Phillips SL, Sherrod SK and Moldenke M. (2005). Soil Biota can change after exotic plant invasion: does this affect ecosystem processes? Ecology, 86: 3007–3017.

Belzile F, Labbé J, LeBlanc MC and Lavoie C. (2009). Seeds contribute strongly to the spread of the invasive genotype of the common reed (*Phragmites australis*). Biological Invasions, 12: 2243-2250.

Bennett MD, Leitch I and Hanson L. (1998). DNA amounts in two samples of angiosperm weeds. Annals of Botany, 82: 121-134.

Bennett MD and Leitch IJ. (2010). Plant DNA C-values Database (release 5.0, December 2010). http://data.kew.org/cvalues/.

Bennett MD. (1972). Nuclear DNA content and minimum generation time in herbaceous plants. Proceedings of the Royal Society B: Biological Sciences, 181: 109–135.

Bennett MD. (1987). Variation in genomic form in plants and its ecological implications. New Phytolologist, 106: 177-200.

Benoit L and Askins R. (1999). Impact of the spread of *Phragmites* on the distribution of birds in Connecticut tidal marshes. Wetlands, 19: 194-208.

Bensch S and Akesson M. (2005). Ten years of AFLP in ecology and evolution: why so few animals? Molecular Ecology, 14: 2899–2914.

Ben-Tal Y and King RW. (1997). Environmental factors involved in coloration of flowers of Kangaroo Paw. Scientia. Horticulture, 72: 35-48.

Bertness MD, Ewanchuk PJ and Silliman BR. (2002). Anthropogenic modification of New England salt marsh landscapes. Proceedings of the National Academy of Sciences of the United States of America, 99:1395-1398.

Bever JD. (2003). Soil community feedback and the coexistence of competitors: conceptual frameworks and empirical tests. New Phytologist, 157: 465–473.

Bickel TO and Closs GP. (2008). Fish distribution and diet in relation to the invasive macrophyte Lagarosiphon major in the littoral zone of Lake Dunstan, New Zealand. Ecology of Freshwater Fish, 17: 10–19.

Blair AC and Hufbauer RA. (2010). Hybridization and invasion: one of North America's most devastating invasive plants shows evidence for a history of interspecific hybridization. Evolutionary Applications, 3: 40– 51.

Blair AC and Wolfe LM. (2004). The evolution of an invasive plant: an experimental study with *Silene latifolia*. Ecology, 85: 3035–3042

Blicker PS, Olson BE and Engel R. (2002). Traits of the invasive *Centaurea maculosa* and two native grasses: effect of N supply. Plant and Soil, 247: 261–269.

Blossey B. (1999). Before, during and after: the need for longterm monitoring in invasive plant species management. Biological Invasions, 1: 301–311.

Bobbink R, Beltman B, Verhoeven JTA and Whigham DF. (2006). Wetlands: Functioning, Biodiversity Conservation, and Restoration. Ecological Studies. Vol. 191. DOI: 10.1007/978-3-540-33189.

Boedeltje G, Bakker JP, Bekker RM, Van Groenendael JM and Soesbergen M. (2003). Plant dispersal in a lowland stream in relation to occurrence and three specific life history traits of the species in the species pool. Journal of Ecology, 91:855–866.

Boers AM and Zedler JB. (2008). Stabilized water levels and *Typha* invasiveness. Wetlands, 28:676-685.

Booth MS, Caldwell MM, Stark JM. (2003). Overlapping resource use in three Great Basin species: implications for community invisibility and vegetation dynamics. Ecology, 91: 36–48.

Bossart JL and Prowell DP. (1998). Genetic estimates of population structure and gene flow: limitations, lessons and new directions. Trends in Ecology and. Evolution, 13: 202–206.

Bossdorf O, Auge H, Lafuma L, Rogers WE, Siemann E and Prati D. (2005). Phenotypic and genetic differentiation between native and introduced plant populations. Oecologia, 144: 1–11.

Brandle M, Stadler J, Klotz S and Brandl R. (2003). Distributional range size of weedy plant species is correlated to germination patterns. Ecology, 84 (1):136–144.

Brisson J, Paradis E and Bellavance ME. (2008). Evidence of sexual reproduction in the invasive common reed (*Phragmites australis* subsp *australis*; Poaceae) in eastern Canada: A possible consequence of global warming? Rhodora, 110: 225-230.

Bronstein JL, Izhaki I, Nathan R *et al.* (2007). Fleshy-fruited plants and frugivores in desert ecosystems. In: Dennis A. J., editor. *Seed Dispersal:* Theory and its Application in a Changing World. Wallingford, UK: CAB International. pp. 148–177.

Broughton LC and Gross KL. (2000). Patterns of diversity in plant and soil microbial communities along a productivity gradient in a Michigan oldfield. Oecologia, 125: 420-427.

Brown JH, Grover JJ, Davidson DW and Lieberman GA. (1975). A preliminary study of seed predation in desert and montane habitats. Ecology, 56 (4): 987–992.

Buckley YM, *et al.* (2006). Management of plant invasions mediated by frugivore interactions. Journal of Applied Ecology, 43: 848—857.

Buggs RJA and Pannell JR. (2007). Ecological differentiation and diploid superiority across a moving ploidy contact zone. Evolution, 61: 125–140.

Bullock JM, Shea K and Skarpaas O. (2006). Measuring plant dispersal: an introduction to field methods and experimental design. Plant Ecology, 186:217-234.

Burdick DM and Konisky RA. (2003). Determinants of expansion for *Phragmites australis*, common reed, in natural and impacted coastal marshes. Estuaries, 26:407-416.

Burke MJW and Grime JP. (1996). An experimental study of plant community invasibility. Ecology, 77: 776–790.

Busch DE and Smith SD. (1995). Mechanisms associated with decline of woody species in riparian ecosystems of the Southwestern US. Ecological Monographs, 65, 347–370.

Byers JE. (2002a). Impact of non-indigenous species on natives enhanced by anthropogenic alteration of selection regimes. Oikos, 97:449–458.

Byers JE. (2002b). Physical habitat attribute mediates biotic resistance to nonindigenous species invasions. Oecologia, 130: 146–156.

Cadotte MW and Lovett–Doust J. (2001). Ecological and taxonomic differences between native and introduced plants of Southwestern Ontario. Ecoscience, 8: 230-238.

Cadotte MW, Murray BR and Lovett-Doust J. (2006). Ecological patterns and biological invasions: using regional species inventories in macroecology. Biological Invasion, 8: 809-821.

Cain SA. (1944). Foundations of Plant Geography.Harper and Brothers, London.

Callaway RM, Thelen GC, Rodringuez A and Holben WE (2004). Soil biota and exotic plant invasion. Nature, 427: 731–733.fungi interaction between the invader *Centaurea maculosa* and North American natives. Ecology, 85: 1062–1071.

Callaway RM and Maron JL. (2006). What have exotic plant invasions taught us over the past 20 years? Trends in Ecology and Evolution, 21(7): 369-374.

Callaway JC and Josselyn MN. (1992). The introduction and spread of smooth cordgrass (*Spartina alterniflora*) in South San Francisco Bay. Estuaries, 15: 218-26.

Callaway RM and Pennings SC. (2000). Facilitation may buffer competitive effects: Indirect and diffuse interactions among salt marsh plants. American Naturalist, 156: 416–424.

Callaway RM and Ridenour WM. (2004). Novel weapons: Invasive success and the evolution of increased competitive ability. Frontiers in Ecology and the Environment, 2:436-443.

Callaway RM, Cipollini D, Barto K, Thelen GC, Hallett SG, Prati D, Stinson K and Klironomos J. (2008). Novel weapons: invasive plant suppresses fungal mutualists in America but not in its native Europe. Ecology, 89: 1043–1055.

Callaway RM, Lauren PW, Alecu D, Robert P, Alexandra RC, Heinz MS. and Maron J. (2011b). Escape from competition: Neighbors reduce *Centaurea stoebe* Performance at home but not away. Ecology, 92: 2208–2213.

Callaway RM, Schaffner U, Thelen GC, Khamraev A, Juginisov T and Maron J. (2011a). Impact of *Acroptilon repens* on co-occurring native plants is greater in the invader's non-native range. Biological Invasions, 14: 1143–1155.

Callaway RM. (1995). Positive interactions among plants. Botanical Review, 61: 306–349.

Canadian Food Inspection Agency. (2008). Invasive Alien Plants in Canada. [Internet] Ottawa, [cited 2010 June 4], Available from http://www.inspection.gc.ca/english/plaveg/invenv/techrpt/summrese.shtml

Carino DA and Daehler CC. (2002). Can inconspicuous legumes facilitate alien grass invasions? Ecography, 25: 33-41.

Casper BB and Jackson RB. (1997). Plant competition underground. Annual Review of Ecology and Systematics, 28: 545–570.

Chambers RM, Meyerson LA and Saltonstall K. (1999). Expansion of *Phragmites australis* into tidal wetlands of North America. Aquatic Botany, 64: 261-273.

Chen GQ, Guo SL and Yin LP. (2010) Applying DNA C-values to evaluate invasiveness of angiosperms: validity and limitation. Biological Invasions, 12: 1335–1348.

Chen ZJ. (2007). Genetic and epigenetic mechanisms for gene expression and phenotypic variation in plant polyploids. Annual Review in Plant Biology, 58: 377–406.

Chilton EW. (1990). Macroinvertebrate communities associated with three aquatic macrophytes (*Ceratophyllum demersum, Myriophyllum spicatum*, and *Vallisneria americana*) in Lake Onalaska, Wisconsin. Journal of Freshwater Ecology, 5: 455–466.

Christen D and Matlock G. (2006). The role of roadside in plant invasions: a demographic approach. Conservation Biology, 20(2): 385-391.

Clegg SM, Degnan SM, Kikkawa J, Moritz C, Estoup A and Owens IPF. (2002). Genetic consequences of sequential founder events by an island-colonizing bird. Proceedings of the National Academy of Sciences, USA, 99: 8127–8132.

Clutton-Brock TH and Harvey PH. (1991). Comparative approaches to investigating adaptation. Pages7-29 in J.R. Krebs and N.B. Davies (eds.), Behavioral Ecology: An Evolutionary Approach. Blackwell scientific publications, Oxford, UK. Community invasibility. Progress in Physical Geography, 30,409-431.

Colautti RI, Grigorovich IA and MacIsaac HJ. (2006). Propagule pressure: a null model for biological invasions. Biological Invasions, 8: 1023–1037.

Comai L. (2005). The advantages and disadvantages of being polyploid. *Nature* Reviews. Genetics, 6: 836–846.

Connell JH. (1971). On the roles of natural enemies in preventing competitive exclusion in some marine animals and in rain forest trees. In: Dynamics of Populations, Proc. Adv. Stud. Inst. Dyn. Numb. Popul., Oosterbeek, 1970, ed. PJ den Boer, GR Gradwell, pp. 298–312. Wageningen, Neth. Centre Agr. Publ. Doc.

Conway WC, Smith LM and Bergan JF. (2002). Potential allelopathic interference by the exotic Chinese tallow tree (*Sapium sebiferum*). American Midland Naturalist, 148: 43–53.

Coops H and Van der Velde G. (1995). Seed dispersal, germination and seedling growth of six helophyte species in relation to water-level zonation. Freshwater Biology, 34:13-20.

Cotto O, Massot M, Ronce O and Clobert J. (2015). Dispersal as a source of variation in age-specific reproductive strategies in a wild population of lizards. Proc. R. Soc. B, 282: 20151741.

Crawley MJ, Harvey PH and Purvis A. (1996). Comparative ecology of the native and alien floras of the British-isles. Philosophical, 351: 1251- 1259.

Crawley MJ. (1986). The population biology of invaders. Philosophical Transactions of the Royal Society of London Series B. Biological Sciences, 314: 711–731.

Crawley MJ. (1987). What makes a community invasible? In: Cray, A. J., Crawley, M. J. and Edwards, P. J. (Eds.), Colonization, Succession and Stability, Blackwell Scientific Publications, Oxford. Pages, 424–453.

Crawley MJ. (1989). Chance and timing in biological invasions. In: J. A. Drake, H. A. Mooney, F. di Castri, R. H. Groves, F. J. Kruger, M. Rejmanek, and M. Williamson. (Eds), Biological invasions: a global perspective. John Wiley and Sons, New York. Pages, 407-424.

Cristescu MEA, Hebert PDN, Witt JDS, MacIsaac HJ and Grigrovich IA. (2001). An invasion history for *Cercopagis pengoi* based on mitochondrial gene sequences. Limnology and Oceanography, 46: 224–229.

Cruzan MB. (1998). Genetic markers in plant evolutionary ecology. Ecology, 79: 400–412.

Cyr H and Downing JA. (1988). The abundance of phytophilous invertebrates on different species of submerged macrophytes. Freshwater Biology, 20: 365– 374.

D'Antonio CM and Mahall BE. (1991). Root profiles and competition between the invasive, exotic perennial, *Carpobrotus edulis*, and two native shrub species in California coastal scrub. American Journal of Botany, 78: 885–894.

Daehler CC and Carino DA. (1999). Threats of invasive plants to the conservation of biodiversity. In: Chou, C. H., Waller, G. R. and Reinhardt, C. (Eds,), In Biodiversity and

Allelopathy: From Organisms to Ecosystems in the Pacific, Taipei: Academic, Pages, 21-27.

Daehler CC. (1998). The taxonomic distribution of invasive angiosperm plants: ecological insights and comparison to agricultural weeds. Biological Conservation, 84: 167–180.

Daehler CC. (2003). Performance comparisons of co-occurring native and alien invasive plants: implications for conservation and restoration. Annual Review of Ecology Evolution and Systematics, 34, 183–211.

Dalling JW and Hubbell SP. (2002). Seed size, growth rate and gap microsite conditions as determinants of recruitment success for pioneer species. Ecology, 90: 557-568.

Darwin C. 1859. The origin of species. London: J. Murray.

Davis MA and Pelsor M. (2001). Experimental support for a resource based mechanistic model of invasibility. Ecology Letters, 4: 421–428.

Davis MA, Grime JP and Thompson K. (2000). Fluctuating resources in plant communities: a general theory of invasibility. Journal of Ecology, 88: 528–534.

Davis MA. (2003). Biotic globalization: does competition from introduced species threaten biodiversity? Journal Bioscience, 53: 481-489.

Davis MA, Wrage KJ, Reich PB, Tjoelker MG, Schaefer T and Muermann C. (1999). Survival, growth, and photosynthesis of tree seedlings competing with herbaceous vegetation along a water-light nitrogen gradient. Plant Ecology, 145:341-350

de la Pena E, de Clercq N, Bonte D, Roiloa S, Rodriguez-Echeverria S and Freitas H. (2010). Plant–soil feedback as a mechanism of invasion by *Carpobrotus edulis.* Biological Invasions, 12: 3637-3648.

Denis, JF. (2011). Invasion dynamics of exotic and native common reed in freshwater wetlands. Masters Thesis.

Delisle F, Lavoie C, Jean M and Lachance D. (2003). Reconstructing the spread of invasive plants: taking into account biases associated with herbarium specimens. Journal of Biogeography, 30:1033-1042.

Detenbeck NE, Galatowitsch SM, Atkinson J and Ball H. (1999). Evaluating perturbations and developing restoration strategies for inland wetlands In the Great Lakes Basin. Wetlands, 19: 789-820.

di Castri F. (1989). History of biological invasions with special emphasis on the old world. Biological Invasions: a Global Perspective. Drake, J. A., Mooney, H. A., di Castri, F., Groves, R.H., Kruger, F.J., Rejmánek, M. and Williamson, M. (Eds.), John Wiley and Sons, New York. Pages, 1–30.

di Castri F. (1990). On invading species and invaded ecosystems: the interplay of historical chance and biological necessity. In: di Castri, F., Hansen, A. J. and Debussche, M. (Eds.), Biological invasions in Europe and the Mediterranean Basin. Kluwer Academic Publishers, Dordrecht, The Netherlands. Pages, 3-16.

Didham RK, Tylianakis JM, Hutchison MA, Ewers RM and Gemmell NJ. (2005). Are invasive species the drivers of ecological change? Trends in Ecology and Evolution, 20: 470–474.

Diez JM, Sullivan JJ, Hulme PE, Edwards G and Duncan RP. (2008). Darwin's naturalization conundrum: dissecting taxonomic patterns of species invasions. Ecology Letters, 11(7):674–681.

Dow BD and Ashley MV. (1996). Microsatellite analysis of seed dispersal and parentage of saplings in bur oak, *Quercus macrocarpa*. Molecular Ecology, 5: 615–627.

Drexler JZ and Bedford BL. (2002). Pathways of nutrient loading and impacts on plant diversity in a New York peatland. Wetlands, 22: 263-281.

Dubcovsky J and Dvorak J. (2007). Genome plasticity a key factor in the success of polyploid wheat under domestication. Science, 316: 1862–1866

Dukes JS and Mooney HA. (1999). Does global change increase the success of biological invaders? Trends in Ecology and Evolution, 14: 135–139.

Duncan JR and Lockwood JL. (2001). Spatial homogenization of the aquatic fauna of Tennessee: extinction and invasion following land use change and habitat alteration. In *Biotic Homogenization* (ed. J.L. Lockwood and M.L. McKinney), Kluwer Academic Publishers, New York pp 245–257.**Duncan RP and Williams PA.** (2002). Ecology-Darwin's naturalization hypothesis challenge. Nature, 417: 608–609.

Durand J, Garnier L, Dajoz I, Mousset S and Veuille M. (2000). Gene flow in a facultative apomictic Poacea, the savanna grass *Hyparrhenia diplandra*. Genetics, 156: 823–831.

Eckert CG, Manicacci D and Barrett SCH. (1996). Genetic drift and founder effect In native versus introduced populations of an invading plant, *Lythrum salicaria* (Lythraceae). Evolution, 50: 1512–1519

Egertson CJ, Kopaska JA and Dowing JA. (2004). A century of change in macrophyte abundance and composition in response to agricultural eutrophication. Hydrobiologia, 524: 145-156.

Ehrendorfer F. (1980). Polyploidy and distribution. In: Lewis WH. ed.Polyploidy: biological relevance. New York: Plenum, 45–60.38.

Ekrt T, Travniček P, Jarolimova V, Vit P and Urfus T. (2009). Genome size and morphology of the *Dryopteris affinis* group in Central Europe. Preslia, 81: 261–280.**Ellner S, Shmida A. (**1981**).** Why are adaptations for long-range seed dispersal rare in desert plants? Oecologia, 51(1):133–144.

Ellstrand NC and Schierenbeck KA. (2000). Hybridization as a stimulus for the evolution in plants? Proceedings of Natural Academy of Sciences, 97:7043-7050.

Elton CS. (1958). The Ecology of Invasions by Animal and Plants. Methuen, London.

Eppinga MB, Rietkerk M, Dekker SC, De Ruiter PC and Van der Putten WH. (2006). Accumulation of local pathogens: a new hypothesis to explain exotic plant invasions. Oikos, 114: 168–176.

Erfmeier A and Bruelheide H. (2011). Maintenance of high genetic diversity during invasion of *Rhododendron ponticum.* International Journal of Plant Science, 172:795–806.

Esselink P, Fresco LFM and Dijkema KS. (2002). Vegetation change in a man-made salt marsh affected by a reduction in both grazing and drainage. Applied Vegetation Science, 5:17-32.

Estoup A and Clegg SM. (2003). Bayesian inferences on the recent island colonization history by the bird *Zosterops lateralis lateralis*. Molecular Ecology, 12: 657–674.

Estoup A, Beaumont M, Sennedot F, Moritz C and Cornuet J-M. (2004). Genetic analysis of complex demographic scenarios: spatially expanding populations of the cane toad, Bufo marinus. Evolution, 58: 2021–2036.

Estoup A, Jarne P and Cornuet JM. (2002). Homoplasy and mutation model at microsatellite loci and their consequences for population genetics analysis. Molecular Ecology, 11: 1591–1604.

Estoup A, Tailliez C, Cornuet JM and Solignac M. (1995). Size homoplasy and mutational processes of interrupted microsatellitesin two bee species, *Apis mellifera* and *Bombus terrestris.*

Estoup A, Wilson IJ, Sullivan C, Cornuet J-M and Moritz C. (2001). Inferring population history from microsatellite and enzyme data in serially introduced cane toads, Bufo marinus. Genetics, 159: 1671–1687.

Fagan WF, Lewis MA, Neubert MG and van den Driessche P. (2002). Invasion theory and biological control. Ecology Letters, 5: 148–157.

Falconer DS and Mackay TFC. (1996). Introduction to Quantitative Genetics, 4th edn. Longman Group Ltd, Edinburgh, UK.

Falinski JB. (1998). Invasive alien plants and vegetation dynamics. Plant Invasions: Ecological mechanisms and Human Responses. In: U. Starfinger, K. Edwards, I. Kowarik and M. Williamson. (Eds.), Backhuys Publishers, Leiden. Pages, 3–21.

Feinsinger P, Tiebout HM and Young BE. (1991). Do tropical bird-pollinated plants exhibit density-dependent interactions, field experiments? Ecology, 72: 1953-1963.

Feng YL, Cao KF and Zhang JL. (2004). Photosynthetic characteristics, dark respiration,and leaf mass per unit area in seedlings of four tropical tree species grown under three irradiances. Photosynthetica, 42: 431–437

Forsyth SA. (2003). Density-dependent seed set in the Halaekala silversword: evidence for Allee effect. Oecologia, 136: 551-557.

Frankham R. (2005). Resolving the genetic paradox in invasive species. Heredity, 94: 385.

Galatowitsch SM, Anderson NO and Ascher PD. (1999). Invasiveness in wetland plants in temperate North America. Wetlands, 19:733-755.

Gallagher RV, Leishman MR, Miller JT, *et al.* (2011). Invasion success of introduced Australian acacias: the role of species' traits and genome size. Diversity and Distributions, 17: 884–897.

Garcia S, Canela MA, Garnatje T, *et al.* (2008). Evolutionary and ecological implications of genome size in the North American endemic sagebrushes and allies (Artemisia, Asteraceae). Biological Journal of the Linnean Society, 94: 631–649.

Gasso N, Pysˇek P, Vila M and Williamson M. (2010). Spreading to a limit: thetime required for a neophyte to reach its maximum range. Diversity and Distributions, 16: 310–311.

Genovesi P and Shine C. (2004). European strategy on invasive alien species. Nature and Environment 137, Council of Europe: Strasbourg, France, 67.

Genton BJ, Shykoff JA and Giraud T. (2005b). High genetic diversity in French invasive populations of common ragweed, *Ambrosia artemisiifolia*, as a result of multiple sources of introduction. Molecular Ecology, 14: 4275–4285.

Gerber S, Mariette S, Streiff R, Bodenes C and Kremer A. (2000). Comparison of microsatellites and amplified fragment length polymorphism markers for parentage analysis. Molecular Ecology, 9: 1037–1048.

Gerlach JD and Rice KJ. (2003). Testing life history correlates of invasiveness using congeneric plant species. Ecological Applications, 13: 167–179.

Gerry AK and Wilson DS. (1995). The influence of initial size on the competitive responses of six plant species. Ecology, 76: 272–279.

Gervais C, Trahan R, Moreno D and Drolet AM. (1993). Le *Phragmites australis* au Québec: distribution géographique, nombres chromosomiques et reproduction. Canadian Journal of Botany, 71:1386-1393.

Giraud T, Pedersen JS and Keller L. (2002). Evolution of supercolonies: the Argentine ants of southern Europe. Proceedings of the National Academy of Sciences, USA, 99: 6075–6079.

Gleeson SK and Tilman D. (1990). Allocation and transient dynamics of succession on poor soils. Ecology, 71: 1144-1155.

Goergen **E and Daehler CC.** (2001). Reproductive ecology of a native Hawaiian grass (*Heteropogon contortus*; Poaceae) versus its invasive alien competitor (*Pennisetum setaceum;* Poaceae). Plant Sciences, 162: 317–326.

Goergen E and Daehler CC. (2002). Factors affecting seedling recruitment in an invasive grass (*Pennisetum setaceum*) and a native grass (*Heteropogon contortus*) in the Hawaiian Islands. Plant Ecology, 161: 147–156.

Goldberg DE and Landa K. (1991). Competitive effect and response: hierarchies and correlated traits in the early stages of competition. Journal of Ecology, 79: 1013-30.

Goldberg DE and Werner PA. (1983). Equivalence of competitors in plant communities: a null hypothesis and a field experimental approach. American journal of Botany, 70: 1098-1104.

Goldberg DE, Rajaniemi T, Gurevitch J and Stewart-Oaten, A. (1999). Empirical approaches to quantifying interaction intensity: competition and facilitation along productivity gradients. Ecology, 80: 1118-1131.

Goldberg DE. (1990). Components of resource competition in plant communities. In: Grace, J.B. and Tilman. D. (Eds.), Perspectives on plant competition. Academic Press, San Diego, Pages, 27–49.

Gollasch S and Nehring S. (2006). National checklist for aquatic alien species in Germany. Aq. Inv, 4: 245-269.

Gollasch S. (2002). Databases on aquatic alien species: North and Mediterranean Seas and non-European Initiatives. In: Leppäkoski E, Gollasch S and Olenin S (eds) Invasive Aquatic Species of Europe. Distribution, Impacts and Management, Kluwer Academic Publishers, Dordrecht. 520-524.

Goodwin BJ, Mc Allister AJ and Fahring L. (1999). Predicting invasiveness of plant species based on biological information. Conservation Biology, 13: 422-426.

Gopal B and Chamanlal T. (1991). Distribution of aquatic macrophytes in polluted water bodies and their bioindicator value. Verhandlungen International Verein Limnology, 24: 2125–2129.

Gopal B and Zutshi DP. (1998). Fifty years of hydrobiological research in India.Hydrobiologia, 384: 267–290.

Grace JB. (1995). On the measurement of plant competition intensity. Ecology, 76: 305-308.

Gratton C and Denno RF. (2006). Arthropod food web restoration following removal of an invasive wetland plant. Ecological Applications, 16: 622-631.

Green D. S. (1980). The terminal velocity and dispersal of spinning samaras. The American Journal of Botany, 67:1218–1224.

Gregory TR. (2005). The evolution of the genome. Amsterdam: Elsevier Academic Press.

Grime JP. (1979). Plant strategies and vegetation processes. John Wiley and Sons Chichester.

Grime JP. (1988). A Functional Approach to Common British Species. Comparative Plant Ecology. Unwin Hyman, London, UK.

Griz LMS and Machado IS. (2001). Fruiting phenology and seed dispersal syndromes in caatinga, a tropica dry forest in northeast of Brazil. Journal of Tropical Ecology, 17:303-321.

Grotkopp E and Rejma´nek M. (2007) High seedling relative growth rate and specific leaf area are traits of invasive species: phylogenetically independent contrasts of woody angiosperms. American Journal of Botany, 94: 526–532.

Grotkopp E, Rejma´nek M and Rostl TL. (2002). Towards a causal explanation of plant invasiveness: seedling growth and life history strategies of 29 pine (Pinus) species. American Naturalist, 159: 396–419.

Grotkopp E, Rejma´nek M, Sanderson MJ and Rost TL. (2004). Evolution of genome size in pines (Pinus) and its life-history correlates: supertree analyses. Evolution, 58: 1705–1729.

Groves RH, Panetta FD and Virtue JG. (2001). Weed Risk Assessment.CSIRO Publishing, Collingwood, Australia.

Gulias J, Flexas J, Mus M, Cifre J, Lefi E and Medrano H. (2003). Relationship between maximum leaf photosynthesis, nitrogen content and specific leaf area in Balearic endemic and non-endemic Mediterranean species. Annals of Botany (London), 92: 215–222.

Guo WH, Wang RQ, Zhou SL, Zhang S, and Zhang ZG. (2003). Genetic diversity and clonal structure of *Phragmites australis* in the Yellow River delta of China. Biochemical Systematics and Ecology, 31: 1093-1109.

Gurevitch J and Padilla DK. (2004). Are invasive species a major cause of extinctions? Trends in Ecology and Evolution, 19: 470-474.

Gurevitch J, Morrow LL, Wallace A and Walsh JS. (1992). A meta analysis of field experiments on competition. American Naturalist, 140: 539-572.

Gutterman Y. (2002). Survival Strategies of Annual Desert Plants. Heidelberg, Germany: Springer.

Haldane JBS. (1948). The theory of a cline. Journal of Genetics, 48: 277–283.

Hamilton MA, Murray BR, Cadotte MW, Hose GC, Baker AC, Harris CJ and Licari D. (2005). Life history correlates of plant invasiveness at regional and continental scales. Ecology Letters, 8:1066 1074.

Hanfling B, Carvalho GR and Brandl R. (2002). mt-DNA sequences and possible invasion pathways of the Chinese mitten crab. Marine Ecology Progress Series, 238: 307–310.

Hanski I. (2001). Population dynamic consequences of dispersal in local populations and in metapopulations. pp. 283–98.

Hanson SR, Osgood DT and Yozzo DJ. (2002). Nekton use of a *Phragmites australis* marsh on the Hudson river, New York, USA. Wetlands, 22:326-337. 71.

Hansen DL, Lambertini C, Jampeetong A and Brix H. (2007). Clone-specific differences in *Phragmites australis*: Effects of ploidy level and geographic origin. Aquatic Botany, 86:269-279.

Harms KE, Wright SJ, Calderon O, Hernandez A and Herre EA. (2000). Pervasive density dependent recruitment enhances seedling diversity in a tropical forest. Nature, 404:493–495.

Harper JL. (1977). Population biology of plants. Academic Press, London. predictors of seed output variation between species. Oikos, 92: 479-490.

Harrison S, Rice K and Maron J. (2001). Habitat patchiness promotes invasion by alien grasses on serpentine soil. Biological conservation, 100: 45-53.

Haslam SM. (1972). *Phragmites-communis* Trin. Journal of Ecology, 60: 585-610.

Havens KJ, Berquist H and Priest NI. (2003). Commun reed grass, *Phragmites australis*, Expansion into constructed wetlands: are we mortgaging our wetland future? Estuaries, 26:417-423.

Hawkes CV, Douglas AE and Fitter AH. (2010). Origin, local experience, and the impact of Biotic interactions on native and introduced *Senecio* species. Biological Invasions, 12: 113–124.

Hayes KR and Barry SC. (2008). Are there any consistent predictors of invasion success? Biological Invasions, 10: 483–506.

Higgins SI and Richardson DM. (1998). Pine invasions in the southern hemisphere: modelling interactions between organism, environment and disturbance. Plant Ecology, 135: 79–93.

Hedrick PW. (1999). Perspective: highly variable loci and their interpretation in evolution and conservation. Evolution, 53:313–318.

Hejda M, Pysˇek P and Jarosˇı´k V. (2009). Impact of invasive plants on the species richness, diversity and composition of invaded communities. Journal of Ecology, 97: 393–403.

Henery ML and Westoby M. (2001). Seed mass and seed nutrient content as predictors of seed output variation between species. Oikos, 92: 479-490.

Hershner C and Havens KJ. (2008). Managing invasive aquatic plants in a changing system: Strategic consideration of ecosystem services. Conservation Biology, 22: 544-550.

Hierro JL, Maron JL and Callaway RM. (2005). A biogeographic approach to plant invasions: the importance of studying exotics in their introduced and native range. Ecology, 93: 5–15.

Hobbs RL and Huenneke LF. (1992). Disturbance, diversity and invasion: implications for conservation. Conservation Biology, 6: 324–337.

Hodgson JG. (1987). Why do so few plant species exploit productive habitats? An investigation into cytology, plant strategies and abundance within a local flora. Functional Ecology, 1: 243–250.

Hoffmann J. (1998). Assessing the effects of environmental changes in a landscape by means of ecological characteristics of plant species. Landscape and Urban Planning, 41: 239–248.

Holt RD. (2009). Darwin, Malthus, and movement: a hidden assumption in the demographic foundations of evolution. - Isreal Journal of Ecolog and Evolution, 55: 189–198.

Hovestadt T, Messner S and Poethke HJ (2000a**).** Evolution of reduced dispersal mortality and 'fat-tailed' in dispersal kernels in autocorrelated landscaps.Proceedings of the Royal Society B,. London 268: 385-91.

Hovestadt T, Poethke HJ and Messner S. (2000b). Variability in dispersal distances generates typical successional patterns: a simple simulation model. Oikos, 90:612–619.

Howard RJ, Travis SE and Sikes BA. (2008). Rapid growth of a Eurasian haplotype of *Phragmites australis* in a restored brackish marsh in Louisiana, USA. Biological Invasions, 10:369-379.

Howard TG and Goldberg DE. (2001). Competitive response hierarchies for germination, growth, and survival and their influence on abundance. Ecology, 82: 979–990.

Howe HF and Smallwood J. (1982). Ecology of seed dispersal. Annual Review of Ecology and systematics, 13:201-228.

Hudon C, Gagnon P and Jean M. (2005). Hydrological factors controlling the spread of common reed (*Phragmites australis*) in the St. Lawrence River (Quebec, Canada). Ecoscience, 12: 347-357.

Huenneke LF and Thomson JK. (1995). Potential interference between a threatened endemic thistle and an invasive nonnative plant. Conservation Biology, 9: 416–25.

Hughes L, Dunlop M, French K, *et al.* (1994). Predicting dispersal spectra: a minimal set of hypotheses based on plant attributes. Journal of Ecology, 82(4):933–950.

Hui C, Richardson DM, Robertson MP, *et al.* (2011). Macroecology meets invasion ecology: linking native distribution of Australian acacias to invasiveness. Diversity and Distributions, 17: 872–883.

Hull-Sanders HM, Johnson RH, Owen HA and Meyer GA. (2009). Effects of polyploidy on secondary chemistry, physiology, and performance of native and invasive genotypes of *Solidago gigantea* (Asteraceae). American Journal of Botany, 96: 762–770.

Hulme PE. (2003). Biological invasions: winning the science battles but losing the conservation war? Oryx, 37, 178-193. https://doi.org/10.1017/S003060530300036X

Hulme PE. (2011). Contrasting impacts of climate driven phenology on changes in alien and native plant species distributions. New Phytologist, 189:272 281.

Hunter KL, Fox DA, Brown LM and Able KW. (2006). Responses of resident marsh fishes to stages of *Phragmites australis* invasion in three mid Atlantic estuaries. Estuaries and Coasts, 29:487-498.

Husband BC and Schemske DW. (1997). The effect of inbreeding in diploid and tetraploid populations of Epilobium angustifolium (Onagraceae): implications for the genetic basis of inbreeding depression. Evolution, 51:737–746.

Husband BC and Barrett SCH. (1991). Colonization history and population genetic structure of *Eichhornia paniculata* in Jamaica. Heredity, 66: 287–296.

Inderjit and Van der Putten WH. (2010). Impacts of soil microbial communities on exotic plant invasions. Trends in Ecology and Evolution, 25: 512-551.

Janzen DH. (1970). Herbivores and number of tree species in tropical forests. The American Naturalist,104: 501–528.

Jodoin Y, Lavoie C, Villeneuve P, Theriault M, Beaulieu J and Belzile F. (2008). Highways as corridors and habitats for the invasive common reed *Phragmites australis* in Quebec, Canada. Journal of Applied Ecology, 45:459-466.

Johnson PT, Olden JD and Vander Zanden MJ. (2008). Dam invaders: impoundments facilitate biological invasions in freshwaters. Frontiers in Ecology and the Environment, 6: 357–363.

Jones DL, Hodge A and Kuzyakov Y. (2004). Plant and mycorrizal regulation of rhizodeposition. New Phytologist, 163: 459-480.

Kak AM. (1990). Aquatic and wetland vegetation of Kashmir Himalaya. J Econ Taxon Bot. 14: 1-14.

Kao-Kniffin J and Balser TC. (2007). Soil fertility and the impact of exotic invasion on microbial communities in Hawaiian forests. Microbial Ecology, 56: 55-63.

Katz GL, Friedman JM and Beatty SW. (2001). Effects of physical disturbance and granivory on establishment of native and alien riparian trees in Colorado U.S.A. Diversity and Distributions, 7: 1–14.

Keane RM and Crawley MJ. (2002). Exotic plant invasions and the enemy release hypothesis. Trends in Ecology and Evolution, 17: 164–170.

Keast A. (1984). The introduced aquatic macrophyte, Myriophyllum spicatum, as habitat for fish and their invertebrate prey. Canadian Journal of Zoology, 62: 1289–1303.

Keller BEM. (2000). Plant diversity in *Lythrum*, *Phragmites* and *Typha* marshes, Massachusetts, U.S.A. Wetlands ecology and management, 8:391-401.

Kettenring KM and Whigham DF. (2009). Seed viability and seed dormancy of non-native *Phragmites australis* in suburbanized and forested watersheds of the Chesapeake Bay, USA. Aquatic Botany, 91:199-204.

King RS, Deluca WV, Whigham DF and Marra PP. (2007). Threshold effects of coastal urbanization on *Phragmites australis* (common reed) abundance and foliar nitrogen in Chesapeake Bay. Estuaries and Coasts, 30:469-481. 72

Kinnison MT, Unwin MJ, Hendry AP and Quinn TP. (2001). Migratory costs and the evolution of egg size and number in introduced and indigenous salmon populations. Evolution, 55: 1656–1667.

Kiviniemi K and Eriksson O. (1999). Dispersal, recruitment and site occupancy of grassland plants in fragmented habitats. Oikos, 86: 241–253.

Klironomos JN. (2002). Feedback with soil biota contributes to plant rarity and invasiveness in communities. Nature, 417: 67–70.

Klotz S. (1990). Species/area and species/inhabitants relations in European cities. Urban ecology plants and plant communities in urban environments. Sukopp, H., Hejny S., and Kowarik, I. (Eds.), Academic Publishing, The Hague, Pages. 99–104.

Knee M and Thomas LC. (2002). Light utilization and competition between *Echinacea purpurea, Panicum virgatum* and *Ratibida pinnata* under greenhouse and field conditions. Ecological Restoration, 17: 591–599.

Knight CA and Beaulieu JM. (2008). Genome size scaling through phenotype space. Annals of Botany, 101: 759–766.

Knight CA, Molinari NA and Petrov DA. (2005). The large genome constraint hypothesis: evolution, ecology and phenotype. Annals of Botany, 95: 177–190.

Knops JMH, Tilman DH, Naeem S, Mitchell CE, Haarstad J, Ritchie ME, Hwe KM, Reich PB, Siemann E and Groth J. (1999). Effects of plant richness on invasion dynamics, disease outbreaks, insect abundance and diversity. Ecology Letters, 2: 286- 293.

Kolar CS and Lodge DM. (2001). Progress in invasion biology: predicting invaders. Trends in Ecology and Evolution, 16: 199–204.

Kolar CS and Lodge DM. (2002). Ecological predictions and risk assessment for alien fishes in North America. Science, 298: 1233–1236.

Kolbe JJ, Glor RE, Rodriguez-Schettino L, Chamizo-Lara A, Larson A and Losos JB. (2004). Genetic variation increases during biological invasion by a Cuban lizard. Nature, 431: 177–181.

Koppitz H. (1999). Analysis of genetic diversity among selected populations of *Phragmites australis* world-wide. Aquatic Botany, 64: 209–221.

Kotanen PM. (1997). Effects of experimental soil disturbance on revegetation by natives and exotics in coastal Californian meadows. Journal of Applied Ecology, 34: 631–644.

Kourtev PS, Ehrenfeld JG and Haggblom M. (2002). Exotic plant species alter the microbial community structure and function in the soil. Ecology, 83: 3152–3166.

Krivanek M, Pysek P and Jarosik V. (2006). Planting history and propagule pressure as predictors of invasion by woody species in a temperate region. Conservation Biology, 20:1487 1498.

Kubatova B, Travnicek P, Bastlova D, Curn V, Jarolimova V and Suda J. (2008). DNA ploidy level variation in native and invasive populations of *Lythrum salicaria* at a large geographical scale. Biogeography, 35: 167-176.

Kubesova M, Moravcová L, Suda J, Jarosík V and Pysek P. (2010). Naturalized plants have smaller genomes than their non-invading relatives: a flow cytometric analysis of the Czech alien flora. – Preslia, 82: 81–96.

Kuhl H, Koppitz H, Rolletschek H and Kohl JG. (1999). Clone specific differences in a *Phragmites australis* stand I. Morphology, genetics and site description. Aquatic Botany, 64: 235-246.

Kulesza AE, Holomuzki JR and Klarer DM. (2008). Benthic community structure in stands of *Typha angustifolia* and herbicide-treated and untreated *Phragmites australis*. Wetlands, 28:40-56.

Kulmatiski A and Beard KH. (2008). Decoupling plant growth from landuse legacies in soil microbial communities. Soil Biology and Biochemistry, 40: 1059-1068.

Kuster EC, Kuhn I, Bruelheide H and Klotz S. (2008). Trait interactionshelp explain invasion success in the German flora. Ecology, 96: 860–868.

Lachmuth S, Durka W and Schurr FM. (2010). The making of a rapid plant invader: genetic diversity and differentiation in the native and invaded range of Senecio inaequidens. Molecular Ecology, 19:3952–3967.

Lafuma L, Balkwill K, Imbert E, Verlaque R and Maurice S. (2003). Ploidy level and origin of the European invasive weed Senecio inaequidens (Asteraceae). Plant Systematics and Evolution, 243: 59–72.

Lake JC and Leishman MR. (2004). Invasion success of exotic plants in natural ecosystems: the role of disturbance, plant attributes and freedom from herbivores. Biological Conservation, 117: 215–226.

Lambers H and Poorter H. (1992). Inherent variation in growth rate between higher plants: a search for physiological causes and ecological consequences. Advances in Ecological Research, 23:188–261.

Lambert AM and Casagrande RA. (2006). Distribution of native and exotic *Phragmites australis* in Rhode island. Northeastern Naturalist, 13: 551-560.

Lambertini C, Gustafsson M, Frydenberg J, Lissner J, Speranza M and Brix H. (2006). A phylogeographic study of the cosmopolitan genus *Phragmites* (Poaceae) based on AFLPs. Plant Systematics and Evolution, 258: 161-182.

Lambrinos JG. (2004). How interactions between ecology and evolution influence contemporary invasion dynamics. Ecology, 85: 2061–2070.

Lande R. (1980). Genetic variation and phenotypic evolution during allopatric speciation. The American Naturalist, 116: 463–479.

Lavergne S, Muenke NJ and Molofsky J. (2010). Genome size reduction can trigger rapid phenotypic evolution in invasive plants. Annals of Botany, 105: 109–116.

Lavoie C, Jean M, Delisle F and Letourneau G. (2003). Exotic plant species of the St Lawrence River wetlands: a spatial and historical analysis. Journal of Biogeography, 30:537-549.

Lavorel S, Prieur-Richard AH. and Grigulis K. (1999). Invasibility and diversity of plant communities: from patterns to processes. Diversity and Distribution, 5: 41-49.

League MT, Colbert EP, Seliskar DM and Gallagher JL. (2006). Rhizome growth dynamics of native and exotic haplotypes of *Phragmites australis* (common reed). Estuaries and Coasts, 29:269-276.

Lee CE. (1999). Rapid and repeated invasions of fresh water by the copepod *Eurytemora affinis*. Evolution, 53: 1423–1434.

Lee CE. (2002). Evolutionary genetics of invasive species. Trends in Ecology and Evolution, 17: 386–391.

Leger EA and Rice KJ. (2003). Invasive California poppies (*Eschscholzia californica* Cham.) grow larger than native individuals under reduced competition. Ecological Letters, 6: 257–264.

Leishman MR and Thomson VP. (2005). Experimental evidence for the effects of additional water, nutrients and physical disturbance on invasive plants in low fertility Hawkesbury Sandstone soils, Sydney, Australia. Journal of Ecology, 93:38–49.

Leishman MR, Haslehurst T, Ares A and Baruch Z. (2007). Leaf trait relationships of native and invasive plants: community- and global-scale comparisons. New Phytology, 176:635–643.

Leishman MR, Westoby M and Jurado E. (1995). Correlates of seed size variation: a comparison among five temperate floras. Journal of Ecology, 83(3): 517–530.

Leitch IJ and Bennett MD. (2004). Genome downsizing in polyploid plants. Biological Journal of the Linnean Society, 82: 651–663.

Leitch IJ and Bennett MD. (2007). Genome size and its uses: the impact of flow cytometry. In: Dolezˇel J, Greilhuber J, Suda J. eds. Flow cytometryn with plant cells. Analysis of genes, chromosomes and genomes. Weinheim: Wiley-VCH, 153–176.

Lelong B, Lavoie C and Thériault M. (2009). Quels sont les facteurs qui facilitent l'implantation du roseau commun (*Phragmites australis*) le long des routes du sud du Québec? Ecoscience, 16:224-237. 73.

Lelong B, Lavoie C, Jodoin Y, and Belzile F. (2007). Expansion pathways of the exotic common reed (*Phragmites australis*): a historical and genetic analysis. Diversity and Distributions, (Diversity Distrib.), 13:430–437.

Lenssen JPM, Menting FBJ, Van der Putten WH and Blom C. (2000). Variation in species composition and species richness within *Phragmites australis* dominated riparian zones. Plant Ecology, 147:137-146.

Lepart J and Debussche M. (1991). Invasion processes as related to succession and disturbance. In: Groves RH and DiCastri F (eds) Biogeography of Mediterranean Invasions, pp 159–178. Cambridge University Press, Cambridge, UK.

Levin DA. (1983). Polyploidy and novelty in flowering plants. American Naturalist, 122: 1–25.

Levin DA. (2000). The origin, expansion and demise of plant species. Oxford: Oxford University Press.

Levin DA. (2003). Ecological speciation: lessons from invasive species. Systematic Botany, 28:643–650.

Levine JM, Vilà M, D'Antonio CM, Dukes JS, Grigulis K and Lavorel S. (2003). Mechanisms underlying the impacts of exotic plant invasions. Proceedings of the Royal Society London B, 270: 775–781.

Lindsay A, Turnbull MR and Crawley MJ. (1999). Seed mass and the competition colonization trade-off: a sowing experiment. Ecology, 87: 899-912.

Liu H and Stiling P. (2006). Testing the enemy release hypothesis: a review and meta-analysis. Biological Invasions, 8:1535-1545.

Lloret F, Médail F, Brundu G and Hulme P. (2004). Local and regional abundance of exotic plant species on Mediterranean islands: are species traits important? Global Ecology and Biogeography, 13:37– 45.

Lloret F, Médail F, Brundu G, Camarda I, Moragues E, Rita J, Lambdon B. and Hulme P. (2005). Species attributes and invasion success by alien plants in Mediterranean islands. Journal of Ecology, 93: 512–520.

Lockwood JL, Cassey P and Blackburn T. (2005). The role of propagule pressure in explaining species invasions. Trends in Ecology and Evolution, 20:223- 228.

Lodge DM, Williams SL, MacIsaac H, Hayes K, Leung B, Reichard S, Mack RN, Moyle PB, Smith M, Andow DA. *et al.* (2006). Biological invasions: recommendations for U.S. policy and management [ESA Position Paper]. Public Affairs Office, Ecological Society of America, Washington, D.C, USA.

Lodge DM. (1993). Biological invasions: lessons from ecology. Trends in Ecology and Evolution, 8:133-136.

Lonsdale WM and Farrell GS. (1998). Testing the effects on *Mimosa pigra* of a biological control agent *Neurostrota gunniella* (Lepidoptera: Gracillaridae), plant competition and fungi under field conditions. Biocontrol Science and Technology, 8: 500-845.

Lorenzo P, Rodriguez-Echeverri S, Luis Gonzalez L and Freitas H. (2010). Effect of invasive *Acacia dealbata* Link on soil microorganisms as determined by PPCR-DGGE. Applied Soil Ecology, 44(3): 245-251.

Lowry E and Lester SE. (2006). The biogeography of plant reproduction: potential determinants of species' range sizes. Journal of Biogeography, 33: 1975–1982.

Lynch M, Pfrender M and Spitze K. (1999). The quantitative and molecular genetic architecture of a subdivided species. Evolution, 53: 100–110.

Lyons J. (1989). Changes in the abundance of small littoral-zone fishes in Lake, Mendota, Wisconsin. Canadian Journal of Zoology, 67: 2910–2916.

Mack RN. (1996). Predicting the identity and fate of plant invaders: Emergent and emerging approaches. Biological Conservation, 78(1-2):107–121.

Maheu-Giroux M and De Blois S. (2007). Landscape ecology of *Phragmites australis* invasion in networks of linear wetlands. Landscape Ecology, 22:285-301.

Mal TK, Lovett-Doust J and Lovett-Doust L. (1997). Time-dependent competitive displacement of *Typha angustifolia* by *Lythrum salicaria.* Oikos, 79: 26-33.

Mandak B, Bimova K, Pysek P, Stepanek J and Plackova I. (2005). Isoenzyme diversity in *Reynoutria* (Polygonaceae) taxa: Escape from sterility by hybridization. Plant Systematics and Evolution, 253: 219-230.

Mandak B, Pysek P, Lysak M, Suda J, Krahulcova A and Bımova K. (2003).Variation in DNA-ploidy levels of *Reynoutria* taxa in the Czech Republic. Annals of Botany, 92: 265–272.

Manel S, Gaggiotti OE and Waples RS. (2005). Assignment methods: matching biological questions with appropriate techniques. Trends in Ecology and Evolution, 20: 136–142.

Manel S, Schwartz MK, Luikart G and Taberlet P. (2003). Landscape genetics: combining landscape ecology and population genetics. Trends in Ecology and Evolution, 18: 189–197.

Mangla S, Inderjit and Callaway RM. (2008). Exotic invasive plant accumulates native soil pathogen which inhibit native plants. Ecology, 96: 58–67.

Maron JL, Vilà M, Bommarco R, Elmendorf S and Beardsley P. (2004). Rapid evolution of an invasive plant. Ecological Monographs, 74: 261–280.

Martens SN, Breshears DD, Meyer CW and Barnes FJ. (1997). Scales of above-ground and below-ground competition in a semi-arid woodland detected from spatial pattern. Vegetation Science, 8: 655– 664.

Mason RAB, Cooke J, Moles AT and Leishman MR. (2008). Reproductive output of invasive versus native plants. Global Journal of Ecology and Biogeography, 17: 633–640.

Masters RA and Sheley RL. (2001). Principles and practices for managing rangeland invasive plants. Journal of Range Management, 54: 502–517.

Matlack G. R. (1987). Diaspore size, shape, and fall, behaviour in wind-dispersed plant species. American Journal of Botany, 1987 (8): 1150-1160.

McDowell SCL. (2002). Photosynthetic characteristics of invasive and noninvasive species of *Rubus* (Rosaceae). American Journal of Botany, 89:1431-1438.

McGlynn CA. (2007). Native and invasive plant interactions in wetlands and the minimal role of invasiveness. Biological Invasions, 11: 1929-1939.

McIntyre S and Lavorel S. (1994a). How environmental and disturbance factors influence species composition in temperate Australian grasslands. Vegetation Science, 5: 373–384.

McKinney ML. (2002). Urbanization, biodiversity and conservation. BioScience, 52: 883–890.

McNeeley JA, Mooney HA, Neville LE, Schei P and Waage JK. (2001). Global Strategy on Invasive Alien Species. IUCN – the World Conservation Union, Gland.

Meadows RE and Saltonstall K. (2009). Distribution of native and introduced *Phragmites australis* in freshwater and oligohaline tidal marshes of the Delmarva peninsula and southern New Jersey. The Journal of the Torrey Botanical Society, 134:99-107.

Meffin R, Miller AL, Hulme PE, Duncan RP (2010). Experimental introduction of the alien weed *Hieracium lepidulum* reveals no significant impact on montane plant communities in New Zealand. Diversity and Distribution, 16: 804-815.

Melgoza G, Nowak RS and Tausch RJ. (1990). Soil water exploitation after fire: competition between *Bromus tectorum* (cheatgrass) and two native species. Oecologia, 83: 7–13.

Meyer DL, Johnson JM and Gill JW. (2001). Comparison of nekton use of *Phragmites australis* and *Spartina alterniflora* marshes in the Chesapeake Bay, USA. Marine Ecology Progress Series, 209: 71-84.

Meyerson LA and Mooney HA. (2007). Invasive alien species in an era of globalization Frontiers in Ecology and Environment, 5(4):199–208.

Meyerson LA, Saltonstall K, Windham L, Kiviat E and Findlay S. (2000). A comparison of *Phragmites australis* in fresh water and brakish marsh environments in North America. Wetlands Ecology Notes, 1:76-78.

Meyerson LA, Viola DV, and Brown RN. (2010). Hybridization of invasive *Phragmites australis* with a native subspecies in North America. Biological Invasions, 12:103-111.

Miller RR, Williams JD and Williams JE. (1989). Extinctions of North American fishes during the past century. Fisheries, 14:(6) 22–38.

Miller TE, Kneitel JM and Burns JH. (2002). Effect of community structure on Invasion success and rate. Ecology, 83: 898–905.

Miller TE and Werner PA. (1987). Competitive effects and responses between plant species in a first-year old-field community. Ecology, 68: 1201–1210.

Minchin D. (2007). Aquaculture and transport in a changing environment: overlap and links in the spread of alien biota. Marine Pollution Bulletin, 55: 302-313.

Minchinton TE and Bertness MD. (2003). Disturbance-mediated competition and the spread of Phragmites australis in a costal marsh. Ecological Applications, 13:1400-1416

Minchinton TE, Simpson JC and Bertness MD. (2006). Mechanisms of exclusion of native coastal marsh plants by an invasive grass. J Ecol. 94: 342-354.

Mitchell CE and Power AG. (2003). Release of invasive plants from fungal and viral pathogens. Nature, 421: 625-627.

Morand ME, Brachet S, Rossignol P, Dufour J and Frascaria-Lacoste N. (2002). A generalized heterozygote deficiency assessed with microsatellites in French common ash populations. Molecular Ecology, 11: 377–385.

Moravcová L, Pyšek P, Jarošík V and Pergl J. (2015). Getting the Right Traits: Reproductive and Dispersal Characteristics Predict the Invasiveness of Herbaceous Plant Species. Plos one, 10 (4): e0123634.

Moravcová L, Pyšek P, Jarosik V, Havlícková V and Zákravský P. (2010). Reproductive characteristics of neophytes in the Czech Republic: traits of invasive and non-invasive species. Preslia, 82: 365-390.

Morgan HD and Westoby M. (2005). The relationship between nuclear DNA content and leaf strategy in seed plants. Annals of Botany, 96: 1321–1330.

Morris K, Boon PI, Raulings EJ and White SD. (2008). Floristic shifts in wetlands: the effects of environmental variables on the interaction between *Phragmites australis* (Common Reed) and *Melaleuca ericifolia* (Swamp Paperbark). Marine and Freshwater Research, 59:187-204.

Morris LL, Walck JL and Hidayati SN. (2002). Growth and reproduction of the invasive *Ligustrum sinense* and native *Forestiera ligustrina* (Oleaceae): Implications for the invasion and persistence of a nonnative shrub. International Journal of Plant Sciences, 163: 1001– 1010.

Moyle PB and Light T. (1996). Biological invasions of fresh water: empirical rules and assembly theory. Biological Conservation, 78: 149–161.

Mozdzer TJ and Zieman JC. (2010). Ecophysiological differences between genetic lineages facilitate the invasion of non-native *Phragmites australis* in North American Atlantic coast wetlands. Journal of Ecology, 98:451-458.

Muller-Scharer H. (1991). The impact of root herbivory as a function of plant density and competition: survival, growth and fecundity of *Centaurea maculosa* in field plots. Ecological Applications, 28: 759-776.

Nagel JM and Griffin KL. (2001). Construction cost and invasive potential: comparing *Lythrum salicaria* (Lythraceae) with co-occurring native species along pond banks. American Journal of Botany, 8: 2252– 2258.

Nagel JM and Griffin KL. (2004). Can gas-exchange characteristics help explain the invasive success of *Lythrum salicaria*? Biological Invasions, 6: 101–111.

Nathan R and Muller-Landau HC. (2000). Spatial patterns of seed dispersal, their determinants and consequences for recruitment. Trends in Ecological and Evolution, 15: 278-285.

National Research Council. (2002). Predicting Invasions of Non-indigenous Plants and Plant Pests. National Academy of Sciences, Washington DC, 198 pp.

Nei M, Maruyama T and Chakraborty R. (1975). Bottleneck effect and genetic-variability in populations. Evolution, 29: 1–10.

Nernberg D and Dale MRT. (1997). Competition of five native prairie grasses with *Bromus inermis* under three moisture regimes. Canadian Journal of Botany, 75: 2140–2145

Neuffer B and Hurka H. (1999). Colonization history and introduction dynamics of *Capsella bursa-pastoris* (Brassicaceae) in North America: isozymes and quantitative traits. Molecular Ecology, 8: 1667–1681.

Nie M, Gao LX, Yan JH, Fu XH, Xiao M, Yang J and Li B. (2010). Population variation of invasive *Spartiana alterniflora* can differentiate bacterial diversity in its rhizosphere. Plant Ecology, 209: 219-226.

Niu H, Liu W, Wan F and Liu B. (2007). An invasive aster (*Ageratina adenophora)* invades and dominates forest understories in China: altered soil microbial communities facilitate the invader and inhibit natives. Plant and Soil, 294: 73–85.

Norton JB, Norton U and Monaco TA. (2007). Mediterranean annual grasses in Western North America: kids in a candy store. Plant and Soil, 298: 1–5.

O'Donnell D and Armbruster P. (2009). Evolutionary differentiation of fitness traits Across multiple geographic scales in aedes albopictus (Diptera: Culicidae). Annals of the Entomological Society of America, 102: 1135–1144.

Olden JD, McCarthy JM, Maxted JT, Fetzer WW and Vander Zanen JM. (2006). The rapid spread of rusty crayfish (Orconectus rusticus) with observations on native crayfish declines in Wisconsin (U.S.A.) over the past 130 years. Biological Invasions, 8, 1621-1628.

Ordonez A and Olff H. (2013). Do alien plant species profit more from high resource supply than natives? Atrait based analysis. glob ecol Biogeogr. Preslia, 81: 173–185.

Orians CM and Ward D. (2010). Evolution of plant defenses in non indigenous environments. Annual Review of Entomology, 55: 439– 459.

Orson RA. (1999). A paleoecological assessment of *Phragmites australis* in New England tidal marshes: Changes in plant community structure during the last few millennia. Biological Invasions, 1: 149-158.

Osborn TC, Pires JC, Birchler JA, *et al.* (2003) Understanding mechanisms of novel gene expression in polyploids. Trends in Genetics, 19: 141–147.

Otto SP and Whitton J. (2000). Polyploid incidence and evolution. Annual Review of Genetics, 34: 401–437.

Otto SP. (2007). The evolutionary consequences of polyploidy. Cell, 131: 452–462.

Packett C and Chambers R. (2006). Distribution and nutrient status of haplotypes of the marsh grass *Phragmites australis* along the Rappahannock River in Virginia. Estuaries and Coasts, 29:1222-1225. 75.

Pandit MK, Pocock MJO and Kunin WE. (2011). Ploidy influences rarity and invasiveness in plants. Journal of Ecology, 99: 1108–1115.

Pandit MK, Tan HTW and Bisht MS. (2006). Polyploidy in invasive plant species of Singapore. Botanical Journal of the Linnean Society, 151: 395–403.

Pandit MK. (2006). Continuing the search for pattern among rare plants: are diploid species more likely to be rare? Evolutionary Ecology Research, 8: 543–552.

Parendes LA and Jones JA. (2000). Role of light availability and dispersal in exotic plant invasion along roads and streams in the H. J. Andrews experimental forest, Oregon. Conservation Biology, 14: 64–75.

Parisod C, Holderegger R and Brochmann C. (2010). Evolutionary consequences of autopolyploidy. New Phytologist, 186: 5–17.

Parker IM and Reichard SH. (1998). Critical issues in invasion biology for conservation science. In: Fiedler, P. L. and Kareiva, P. M. (Eds.), Conservation biology for the coming decade. Chapman and Hall, pp. 283-305.

Parsons KC. (2003). Reproductive success of wading birds using *Phragmites* marsh and upland nesting habitats. Estuaries, 26:596-601.

Paschke MW, McLendon T and Redente EF. (2000). Nitrogen availability and oldfield succession in a shortgrass steppe. Ecosystems, 3: 144–158.

Paterson E, Gebbing T, Abel C, Sim A and Telfer G. (2007). Rhizodeposition shapes rhizosphere microbial community structure in organic soil. New Phytologist, 173: 600-610.

Pattison RR, G Goldstein and A Ares. (1998). Growth, biomass allocation and photosynthesis of invasive and native Hawaiian rainforest species. Oecologia, 117:449–459.

Paul J, Kirk H, Freeland J. (2010a). Genetic diversity and differentiation of fragmented reed beds (*Phragmites australis*) in the United Kingdom. Hydrobiologia, 665: 107-115.

Paul J, Vachon N, Garroway C and Freeland J. (2010b). Molecular data provide strong evidence of natural hybridization between native and introduced lineages of *Phragmites australis* in North America. Biological Invasions, 12: 2967–2973

Paun O, Bateman RM, Fay MF, Hedren M, Civeyrel L and Chase MW. (2010). Stable epigenetic effects impact adaptation in allopolyploid orchids (Dactylorhiza: Orchidaceae). Molecular Biology and Evolution, 27:2465–2473.

Paun O, Bateman RM, Fay MF, Luna JA, Moat J, Hedren M and Chase MW. (2011). Altered gene expression and ecological divergence in sibling allopolyploids of Dactylorhiza (Orchidaceae). BMC Evolutionary Biology, 11: 113.

Peakall R, Gilmore S, Keys W, Morgante M and Rafalski A. (1998). Cross species amplification of soybean (Glycine max) simple sequence repeats (SSRs) within the genus and other legume genera: implications for the transferability of SSRs in plants. Molecular Biology and Evolution, 15: 1275–1287.

Pearse DE and Crandall KA. (2004). Beyond FST: analysis of population genetic data for conservation. ConservationGenetics, 5: 585–602.

Perrins J, Williamson M, and Fitter A. (1992). Do annual weeds have predictable characters? Acta Oecologia, 13: 517-533.

Petryna L, Moora M, Nunes CO, Cantero JJ and Zobel M. (2002). Are invaders disturbance limited? Conservation of mountain grasslands in Central Argentina. Applied Vegitation Science, 5: 195–202.

Philipp KR and Field RT. (2005). *Phragmites australis* expansion in Delaware Bay salt marshes. Ecological Engineering, 25: 275-291.

Phillips JA and Price IR. (2002). How different is Mediterranean *Caulerpa taxifolia* (Caulerpales: Chlorophyta) to other populations ofthe species? Marine Ecology Progress Series, 238: 61–71.

Platt WJ and Weis IM. (1977). Resource partitioning and competition within a guild of fugitive prairie plants. American Naturalist. 111:479–513.

Posey MH, Alphin TD, Meyer DL and Johnson JM. (2003). Benthic communities of common reed *Phragmites australis* and marsh cordgrass *Spartina alterniflora* marshes in Chesapeake Bay. Marine Ecology Progress Series, 251: 51-61.

Pringle A, Bever JD, Gardes M, Parrent JL, Rillig MC and Klironomos JN. (2009). Mycorrizal symbioses and plant invasions. Annual Review Ecology, Evolution and Systematics, 40: 699-715.

Proches Ş, Wilson JRU, Richardson DM and Chown SL. (2008). Herbivores, but not other insects, are scarce on alien plants. Austral Ecology, 33:691-700.

Pysek P and Richardson DM. (2007). Traits associated with invasiveness in alian plants. Journal of Biogeography, 33:2040-2050.

Pysek P, Jarosık V, Pergl J, Rindall R, Chytry M, Kuhn I, Tichy L, Danilelka J, Chrtek Jun and Sadlo J. (2009). The global invasionsuccess of Central European plants is related to distributioncharacteristics in their native range and species traits. Diversity and Distributions, 15: 891–903.

Pysek P, Prach K and Smilauer P. (1995). Relating invasion success to plant traits: an analysis of the Czech alien flora. In: Pysek, P., Pranch, K., Rejmanek, M. and Wade, M. (Eds.), Plant invasions: general aspects and special problems. SPB Academic, Amsterdam. Pages, 39-60

Pysek P, Richardson DM and Williamson M. (2004). Predicting and explaining plant invasions through analysis of source area floras: some critical considerations. Diversity and Distributions, 10:179–187.

Pysek P, Sadlo J and Mandak B. (2002). Catalogue of alien plants of the Czech Republic. Preslia, 74: 97-186.

Pysek P, Sadlo J, Mandak B and Jarosik V. (2003). Czech alien flora and a historical pattern of its formation: what came first to Central Europe? Oecologia, 135: 122-130.

Pysek P. (1998). Is there a taxonomic pattern to plant invasions? Oikos, 82: 282–294.

Pysek P and Jarosık V. (2005). Residence time determines the distribution of alien plants. Inderjit. (Ed.), Invasive plants: ecological and agriculturalaspects. Birkhauser, Basel, Pages, 77-96.

Pysek P. (1998). Is there a taxonomic pattern to plant invasions? Oikos, 82: 282–294.

Queller DC, Strassmann JE and Hughes CR. (1993). Microsatellites and kinship. Trends in Ecology and Evolution, 8: 285–288.

Radford IJ and Cousens RD. (2000). Invasiveness and comparative life history traits of exotic and indigenous *Senecio* species in Australia. Oecologia, 125: 531–542.

Radosevich SR, Stubbs MM and Ghersa CM. (2003). Plant invasions – process and patterns. Weed Science, 51: 254–259.

Rahel F. (2002). Homogenization of freshwater faunas. Annual Review of Ecology and Systematics, 33: 291–315.

Rajmanek M and Richardson DM. (1996). what attributes make some plants more invasive? Ecology, 77:1655 1661.

Rajmanek M, Richardson DM and Pysek P. (2005). Plant invasions and invisibility of plant communities. In: Van der Maarel E, editor. Vegetation ecology. Oxford: Blackwell Science, 332-355.

Ramsey J and Schemske DW. (1998). Pathways, mechanisms, and rates of polyploidy formation in flowering plants. Annual Review of Ecology and Systematics, 29: 467–501.

Ravit B, Ehrenfeld JG, Haggblom MM and Bartels M. (2007). The effects of drainage and nitrogen enrichment on *Phragmites australis*, *Spartina alterniflora*, and their root-associated microbial communities. Wetlands, 27:915-927.

Reader RJ and Bonser SP. (1993). Control of plant frequency on an environmental gradient: effects of abiotic variables, neighbours, and predators on *Poa pratensis* and *Poa compressa* (Gramineae). Canadian Journal of Botany, 71: 592–597.

Rehage J S. (2003). Traits underlying invasiveness: a comparison of widespread and endemic species in the genus gambusia (poeciliidae). University of kentucky doctoral dissertations. Paper 262.http://uknowledge.uky.edu/gradschool_diss/262.

Reinhart KO and Callaway RM. (2006). Soil biota and invasive plants. New Phytologist, 170: 445–457.

Reich PB, Walters MB and Ellsworth DS. (1997). From tropics to tundra: global convergence in plant functioning. Proceedings of the National Academy of Sciences, 94: 13730–13734.

Reichman OJ (1979). Desert granivore foraging and its impact on seed densities and distributions. Ecology. 60:1085–1092. doi: 10.2307/1936954.

Rejmanek M, Richardson DM and Pysek P. (2005). Plant invasions and invisibility of plant communities. In: Van der Maarel E. ed. Vegetation ecology. Oxford: Blackwell Publishing, 332–355.

Rejmanek M, Richardson DM, Higgins SI, Pitcairn MJ and Grotkopp E. (2005). Ecology of invasive plants: state of the art. In: Mooney HA, Mack RN, McNeely JA, Neville LE, Schei PJ, Waage JK, eds. Invasive alien species: a new synthesis. Washington, DC, USA: Island Press, 104–161.

Rejmanek M (1999). Invasive plant species and invasible ecosystems. In: O. T. Sandlund, P. J. Schei, and A. Vilken. (Eds.), Invasive Species and Biodiversity Management, Kluwer, Dordrecht. Pages, 79–102.

Rejmanek M and Richardson DM. (1996). What makes some plant species more invasive? Ecology, 77: 1655-1661.

Rejmanek M. (1989). Invasibility of plant communities. In: Drake, J. A., Mooney, H. A., di Castri, F., Groves, R. H., Kruger, F. J., Rejmanek, M. and Williamson, M. A. (Eds.), Biological Invasions. Global Perspective. Wiley and Sons, Chichester, England, 369-388.

Rejmanek M. (1995). What makes a species invasive? In Plant Invasions. Pysek, K., Prach., M, Rejmnek, and Wade. P.M. (Eds.), The Hague, Netherlands: SPB Academy.Pages, 1-11.

Rejmanek M. (1996). A theory of seed plant invasiveness: the first sketch. Biological Conservation, 78: 171–181.

Rejmanek M. (2000). Invasive plants: Approaches and predictions. Australian journal of Ecology, 25: 497-506.

Reshi Z, Shah MA, Rashid I and Khuroo AA. (2008). Alien plant invasions: Threat to wetland health and public wealth. ENVIS Newsletter Sarovar Saurabh, 4: 2-5.

Ricciardi A. (2004). Assessing species invasions as a cause of extinction Trends in Ecology and Evolution, 19: 619.

Ricciardi A. (2007). Are modern biological invasions an unprecedented form of global change? Conservation Biology, 21: 329–336.

Rice D, Rooth J, and Stevenson, JC. (2000). Colonization and expansion of *Phragmites australis* in upper Chesapeake Bay tidal marshes. Wetlands, 20:280-299.

Richardson DM and Pysek P. (2006). Plant invasions: merging the concepts of species invasiveness and community invasibility. Progress in Physical Geography, 30: 409–431.

Richardson DM and Rejmanek M. (2004). Conifers as invasive aliens: a global survey and predictive framework. Diversity and Distributions, 10: 321–331.

Richardson DM, Allsopp N, D'Antonio CM, Milton SJ and Rejmanek M. (2000). Plant invasions – the role of mutualisms. Biological Reviews, 75: 65–93.

Richburg JA, Patterson WA and Lowenstein F. (2001). Effects of road salt and *Phragmites australis* invasion on the vegetation of a western Massachusetts calcareous lake-basin fen. Wetlands, 21:247-255.

Rickey MA and Anderson RC. (2004). Effects of nitrogen addition on the invasive grass *Phragmites australis* and a native competitor *Spartina pectinata*. Journal of Applied Ecology, 41: 888-896.

Ridenour WM and RM Callaway. (2001). The relative importance of allelopathy in interference: the effects of an invasive weed on a native bunchgrass. Oecologia, 126:444–450.**Ridley HN**. (1930). The dispersal of plants throughout the world. L. Reeve & Co., Ltd., Ashford, Kent. 744 p.

Roberts KJ and Anderson RC. (2001). Effect of garlic mustard *Alliaria petiolata* (Beib. Cavara and Grande) extracts on plants and arbuscular mycorrhizal (AM) fungi. American Midland Naturalist, 146: 146–152.

Robertson TL and Weis JS. (2005). A comparasion of epifaunal communities associated with the stems of salt marsh grasses *Phragmites australis* and *Spartina alterniflora*. Wetlands, 25:1-7.

Robinson GR, Quinn JF and Stanton ML. (1995). Invasibility of experimental Habitat Island in a California winter annual grassland. Ecology, 76: 786–794.

Roose ML and Gottlieb DLD. (1976). Genetic and biochemical consequences of polyploidy in Tragopogon. Evolution, 30: 818–830.

Roy J. (1990). In search of the characteristics of plant invaders. In: di Castri, F., Hansen, A. J. and Debussche, M. (Eds.), Biological invasions in Europe and the Mediterranean basin. Kluwer Academic Publishers. Pages, 333-352.

Röder MS, Plashke J, König SU, *et al.* (1995) Abundance, variability, and chromosomal location of microsatellites in wheat. Molecular and General Genetics, 246: 327–333

Rudrappa T, Choi YS, Levia DF, Legates DR, Lee KH and Bais HP. (2009). *Phragmites australis* root secreted phytotoxin undergoes photo-degradation to execute severe phytotoxicity. Plant Signal Behavior, 4:506-513.

Sakai AK, Allendorf FW, Holt JS, Lodge DM, Molofsky J, With KA, Baughman S, Cabin RJ, Cohen JE, Ellstrand NC, McCauley DE, O'Neil P, Parker IM, Thompson JN and Weller SG. (2001). The population biology of invasive species. Annual Review of Ecology and Systematics, 32: 305–332.

Salisbury E. (1961). Weeds and Aliens. Collins Publishing, London.

Saltonstall K and Stevenson JC. (2007). The effect of nutrients on seedling growth of native and introduced *Phragmites australis*. Aquatic Botany, 86:331-336.

Saltonstall K, Peterson PM and Soreng RJ. (2004). Recognition of *Phragmites australis* subsp. *americanus* (*poaceae*: *arundinoideae*) in North America: evidence from morphological and genetic analyses. Swedish International Development Agency 21, 683-692.

Saltonstall K. (2002). Cryptic invasion by a non-native genotype of *Phragmites australis* into North America. Proceedings of theNational Academy of Sciences of the USA, 99: 2445–2449.

Saltonstall K. (2003a). Microsatellite variation within and among North American lineages of *Phragmites australis*. Molecular Ecology, 12:1689-1702.

Saltonstall K. (2003b). Genetic variation among North American populations of *Phragmites australis* : implications for management. Esturies, 26:444-451.

Samson D. A, Philippi T. E, Davidson D. W. (1992). Granivory and competition as determinants of annual plant diversity in the Chihuahuan Desert. Oikos. 65(1):61–80.

Sanon A, Beguiristian T, Cebron A, Berthelin J, Sylla SN and Duponnois R. (2012). Differences in nutrient availability and mycorrizal infectivity in soils invaded by an exotic plant negatively influence the development of indigenous *Acacia* species. Environmental Management 275-279.

Schaefer H, Hardy OJ, Silva L, Barraclough TG and Savolainen V. (2011).Testing Darwin's naturalization hypothesis in the Azores. Ecology Letters, 14: 389–396.

Schierenbeck KA and Ellstrand NC. (2009). Hybridization and the evolution of invasiveness in plants and other organisms. Biological Invasions, 11: 1093–1105.

Schippers P and Olff H. (2000). Biomass partitioning, architecture and turnover of six herbaceous species from habitats with different nutrient supply. Plant Ecology, 149: 219-231.

Schlaepfer DR, Edwards P and Billeter R. (2010). Why only tetraploid *Solidago gigantean* (Asteraceae) became invasive: a common gardencomparison of ploidy levels. Oecologia, 163: 661–673.

Schlotterer C. (2004). The evolution of molecular markers – just amatter of fashion? Nat. Rev. Genet, 5: 63–69.

Schwinning S and Weiner J. (1998). Mechanisms determining the degree of size asymmetry in competition among plants. Oecologia, 113: 447–455.

Scott MC. (2006). Winners and losers among stream fishes in relation to land use legacies and urban development in the southeastern US. Biological Conservation, 127: 301–309.

Scott JK. and F. D. Panetta. (1993). Predicting the Australian weed status of southern African plants. Biogeography, 20: 87-93.

Scott L, Congdon B and Playford J. (1998). Molecular evidence that fireweed (*Senecio madagascariensis,* Asteraceae) is of South African origin. Plant Systematics and Evolution, 213: 251–257.

Seabloom EA, Borer ET, Boucher VL, Burton RS, Cottingham KL, Goldwasser L. *et al.* (2003). Competition, seed limitation, disturbance and re-establishment of California native annual forbs. Ecological applications, 13, 592.

Seastedt TR and Suding KN. (2007). Biotic constraints on the invasion of diffuse Knapweed (*Centaurea diffusa*) in North American grasslands. Oecologia, 151: 626–636.

Seifert EK, Bever JD and Maron JL. (2009). Evidence for the evolution of reduced mycorrhizal dependence during plant invasion. Ecology, 90: 1055–1062.

Serrano-Garcia H, Sans FX and Escarre J. (2007). Interspecific competition between alien and native congeneric species. Acta Oecologica 31: 69 – 78.

Shah MA and Z. Reshi. (2014). Characterization of alien aquatic flora of Kashmir Himalaya: implications for invasion management. Tropical Ecology, 55 (2): 143-157, 2014.

Shay JM and Shay CT. (1986). Prairie marshes in western Canada, with specific reference to the ecology of five emergent macrophytes. Canadian Journal of Botany, 64:443-454.

Shay JM, De Geus PMJ and Kapinga MRM. (1999). Changes in shoreline vegetation over a 50-year period in the Delta marsh, Manitoba in response to water levels. Wetlands, 19:413-425.

Silliman BR and Bertness MD. (2004). Shoreline development drives invasion of *Phragmites australis* and the loss of plant diversity on New England salt marshes. Conservation Biology, 18:1424-1434.

Simberloff D. (2009). The role of Propagule pressure in biological invasions. Annual Review of Ecology, 40:81 102.

Slatkin M. (1973). Gene flow and selection in a cline. Genetics, 75: 733–756.

Smith C. (2008). Invasive Exotic Plants of North Carolina. N.C. Department of Transportation. Raleigh, NC.

Smith MD and Knapp AK. (2001). Physiological and morphological traits of exotic, invasive exotic and native plant species in tall grass prairie. Plant Science, 162(4): 785-792.

Soltis DE and Soltis PS. (1999). Polyploidy: recurrent formation and genome evolution. Trends in Ecology and Evolution, 14: 348–352.

Soltis DE, Buggs RJA, Doyle JJ and Soltis PS. (2010). What we still don't know about polyploidy. Taxon, 59: 1387–1403.

Soltis DE, Soltis PS, Schemske DW, *et al.* (2007). Autopolyploidy in angiosperms: have we grossly underestimated the number of species? Taxon, 56: 13–30.

Soltis PS and Soltis DE. (2000). The role of genetic and genomic attributes in the success of polyploids. Proceedings of the National Academy of Sciences of the United States of America, 97:7051–7057.

Sorensen A. E. (1986). Seed dispersal by adhesion. Annual Review of Ecology and Systematic, 17:443–463.

Spitze K. (1993). Population structure in Daphnia obtusa: quantitative genetics and allozyme variation. Genetics, 135: 367–374.

Squirell J, Hollingsworth PM, Bateman RM *et al.* (2001). Partitioningand diversity of nuclear and organelle markers in native and introduced populations of *Epipactis helleborine* (Orchidaceae). American Journal of Botany, 88: 1409–1418.

Srikwan S, Hufford K, Eggert L and Woodruff DS. (2002). Variable Microsatellite markers for genotyping Tree Shrews, Tupaia, and their potential use in genetic studies of fragmented populations. Science Asia, 28: 93-97.

Stanton LE. (2005). The establishment, expension and ecosystem effects of *Phragmites australis*, an invasive species in coastal Louisiana. In Oceanography and Coastal Sciences (Baton Rouge: Louisiana State University), pp. 166. 77

Stebbins GL and Dawe JC. (1987). Polyploidy and distribution in the European flora: a reappraisal. Botanische Jahrbu¨cher fu¨r Systematik, Pflanzengeschichte und Pflanzengeographie, 108: 343–354.

Stebbins GL. (1950). Variation and Evolution in Plants. Columbia Univ. Press, New York.

Stebbins GL. (1985). Polyploidy, hybridization, and the invasion of new habitats. Annals of the Missouri Botanical Garden, 72: 824–832.

Stinson KA, Campbell SA, Powell JR, Wolfe BE, Callaway RM, Thelen GC, Hallett SG, Prati, D and Klironomos JN. (2006). Invasive plant suppresses the growth of native tree seedlings by disrupting belowground mutualisms. Public Library of Science Biology, 4(5): 140-145.

Stockwell CA, Hendry AP and Kinnison MT. (2003). Contemporary evolution meets conservation biology. Trends in Ecology and Evolution, 18: 94–101.

Stohlgren TJ and Schnase JL. (2006). Risk analysis for biological hazards: what we need to know about invasive species. Risk Assessment, 26: 163–173.

Stohlgren TJ, Schell LD and Vanden Heuvel B. (1999). How grazing and soil quality affect native and exotic plant diversity in rocky mountain grasslands. Ecological Applications, 9: 45-64.

Stoll P and Weiner J. (2000). Neighbourhood view of interactions among individual plants. In: Dieckmann, U., Law, R., Metz, J.A.J. (Eds.), The Geometry of Ecological Interactions. Cambridge University Press, Cambridge, Pages 11–27.

Strong D, Lawton J and Southwood R. (1984). Insects on Plants: Community Patterns and Mechanisms. Blackwell Scientific, Oxford.

Suda J and Pysˇek P. (2010). Flow cytometry in botanical research: introduction. Preslia, 82: 1–2.

Sukopp H. (1998). On the study of anthropogenic plant migrations in central Europe.In: Starfinger. U., Edwards, K. R., Kowarik, I. and Williamson, M. (Eds.), Plant invasions: ecological mechanisms and human responses. Backhuys Publishers, Leiden. Pages, 43–56.

Sun Z-K and He W-M. (2010). Evidence for enhanced mutualism hypothesis: *Solidago canadensis* plants from regular soils perform better. PLoS 5(11): e15418.

Sunnucks P. (2000). Efficient genetic markers for population biology. Trends in Ecology and Evolution, 15: 199–203.

Taberlet P, Waits LP and Luikart G. (1999). Noninvasive genetic sampling: look before you leap. Trends in Ecology and Evolution, 14: 323–327.

te Beest M, Le Roux JJ, Richardson DM , Brysting AK, Jan Suda Kubesova M and Pysek P. (2012). The more the better? The role of polyploidy in facilitating plant invasions. Annals of Botany, 109: 19–45, 2012.

Tewksbury L, Casagrande R, Blossey B, Häfliger P and Schwarzländer M. (2002). Potential for biological control of *Phragmites australis* in North America. Biological Control, 23:191-212.

Thebaud C and Simberloff D. (2001). Are plants really larger in their introduced ranges? American Naturalist, 157: 231-236.

Theoharides KA and Dukes JS. (2007). Plant invasion across space and time: factors affecting non-indigenous species success during four stages of invasion. New Phytologist, 176: 256-273.

Thompson K, Band SR and Hodgson JG. (1993). Seed size and shape predict persistence in soil. Functional Ecology, 7: 236-241.

Thompson, JD and Lumaret R. (1992). The evolutionary dynamics of polyploid plants - origins, establishment and persistence. Trends in Ecology and Evolution, **7:** 302-307.

Thompson K and Grime JP. (1979). Seasonal variation in the seed banks of Herbaceous species in ten contrasting habitats. Ecology, 67: 893-921.

Thompson K, Hodgson JG and Rich TCG. (1995). Native and alien invasive plants: more of the same? Ecography, 18: 390-402.

Thuiller W, Richardson DM, Pysek P, Midgley GF, Hughes GO and Rouget M. (2005). Niche-based modelling as a tool for predicting the risk of alien plant invasions at a global scale. Global Change Biology, 11: 2234–2250.

Tiffney BH. (1984). Seed size, dispersal syndromes and the rise of the angiosperms: evidence and hypothesis. Ann Annals of the Missouri Botanical Garden, 71: 551-576.

Tilman D. (1988). Plant Strategies and the Dynamics and Structure of Plant Communities. Princeton University Press, Princeton, N. J.

Townsend Peterson A. (2003). Predicting the geography of species' invasions via ecological niche modeling. Quaterly Review of Biology, 78: 419– 433.

Traveset A, Ruben H and Manuel N. (2014). The Ecology of Seed Dispersal. CAB International 2014. In: Seeds. The Ecology of Regeneration in Plant Communities, 62 3rd Edition (ed. R.S. Gallagher).

Trebitz AS and Taylor DL. (2007). Exotic and invasive aquatic plants in great lakes coastal wetlands: Distribution and relation to watershed land use and plant richness and cover. Journal of Great Lakes Research, 33:705-721.

Treier UA, Broennimann O, Normand S, Guisan A, Schaffner U, Steinger T and Muller Schaer H. (2009). Shift in cytotype frequency and niche space in the invasive plant *Centaurea maculosa.* Ecology, 90: 1366 – 1377.

Tulbure MG, Johnston CA and Auger DL. (2007). Rapid invasion of a Great Lakes coastal wetland by non-native *Phragmites australis* and *Typha.* Journal of Great Lakes Research, 33:269-279.

Tyser RW and Worley CA. (1992). Alien flora in grasslands adjacent to road and Trail corridors in Glacier National Park, Montana (USA). Conservation Biology, 6: 253–262.

Van Breeman AMM. (1984). Comparative germination ecology of three short-lived monocarpic Boraginaceae. Acta –Botanische-Neerlandica, 33(3): 283-305.

Van der Pijl L. (1972). Principles of dispersal in higher plants. 2 ed. Springer-Verlag, Wu¨ rzburg.

Van der Pijl L. (1982). Principles of Dispersal in Higher Plants. Springer-Verlag, Berlin.

Van Grunsven RHA, van der Putten WH, Bezemer TM, Berendse F and Veenendaal EM. (2010). Plant–soil interactions in the expansion and native range of a poleward shifting plant species. Global Change Biology, 16: 380–385.

Van Kleunen M, Johnson SD and Fischer M. (2007). Predicting naturalization of southern African Iridaceae in other regions. Journal of Applied Ecology, 44: 594–603.

Van Kleunen M, Weber E and Fischer M. (2010). A meta-analysis of trait differences between invasive and non-invasive plant species. Ecology Letters, 13: 235–245.

Van Oppen MJH, Rico C, Turner GF and Hewitt GM. (2000). Extensive homoplasy, nonstepwise mutations, and shared ancestral polymorphism at a complex microsatellite locus in lake Malawicichlids. Molecular Biology and Evolution, 17: 489–498.

Van Oudtshoorn KVR and Van Rooyen MW. (1999). Dispersal Biology of Desert Plants. Berlin, Germany: Springer.

Vasquez EA, Glenn EP, Brown JJ, Guntenspergen GR and Nelson SG. (2005). Salt tolerance underlies the cryptic invasion of North American salt marshes by an introduced haplotype of the common reed *Phragmites australis* (Poaceae). Marine Ecology- Progress Series, 298: 1-8.

Veilleux RE and Lauer FI. (1981). Variation for 2n pollen production in clones of *Solanum phureja* Juz. and Buk. Theoretical and Applied Genetics, 59: 95–100.

Verlaque R, Aboucaya A and Fridlender A. (2002). Invasive alien flora of France: ecology, life-forms and polyploidy. Botanica Helvetica, 112: 121–136.

Viard F, Franck P, Dubois MP, Estoup A and Jarne P. (1998). Variation of microsatellite size homoplasy across electromorphs, loci, and populations in three invertebrate species. Journal of Molecular Ecology and Evolution, 47: 42–51.

Vila M and Weiner J. (2004). Are invasive plant species better competitors than native plant species? Evidence from pairwise experiments. Oikos, 105: 229–238.

Vila M, Basnou C, Pysˇek P, *et al.* (2010). How well do we understand the impacts of alien species on ecosystem services? A pan-European, crosstaxa assessment. Frontiers in Ecology and the Environment, 8: 135–144.

Vila M, Espinar JL, Hejda M, *et al.* (2011). Ecological impacts of invasive alien plants: a meta-analysis of their effects on species, communities and ecosystems. Ecology Letters, 14: 702–708.

Vinogradov AE. (2003). Selfish DNA is maladaptive: evidence from the plant Red List. Trends in Genetics, 19: 609–614.

Vogelsang KM and Bever JD. (2009) Mycorrhizal densities decline in association with non-native plants and contribute to plant invasion. Ecology, 90: 399–407.

Von Holle B and Simberloff D. (2005). Ecological resistance to biological invasion overwhelmed by propagule pressure. Ecological Society of America. https://doi.org/10.1890/05-0427.

Wang J, Tian L, Lee HS, Jiang H, Waston B, Madlung A, Osborn TC, Doerge RW, Comai L and Jeffrey Chen Z. (2006). Genome-wide non additive gene regulation in Arabidopsis allotetraploids. Genetics, 172: 507–517.

Wani GA, Shah MA, Reshi ZA, Dar MA. 2018. Polyploidy determines the stage of invasion; clues from Kashmir Himalayan aquatic flora. Acta Physiolgy Plantarium, https://doi.org/10.1007/s11738-018-2629-4.

Warren RS, Fell PE, Grimsby JL, Buck EL, Rilling GC and Fertik RA. (2001). Rates, patterns, and impacts of *Phragmites australis* expansion and effects of experimental *Phragmites* control on vegetation, macroinvertebrates, and fish within tidelands of the lower Connecticut river. Estuaries, 24:90-107.

Weiner J. (1990). Asymmetric competition in plant populations. Trends in Ecology and Evolution, 5: 360–364.

Weinstein MP and Balletto JH. (1999). Does the common reed, *Phragmites australis*, affect essential fish habitat? Estuaries, 22: 793-802.

Weis JS and Weis P. (2003). Is the invasion of the common reed, *Phragmites australis*, into tidal marshes of the eastern US an ecological disaster? Marine Pollution Bulletin, 46:816-820.

Werner PA and Platt WJ. (1976). Ecological relationships of co-occurring goldenrods (Solidago: compositae) The American Naturalist. 110:959–971.

Westoby M, Falster DS, Moles AT, Vesk PA and Wright IJ. (2002). Plant ecological strategies: some leading dimensions of variation between species. Annual Review of Ecology and Systematics, 33: 125–159.

Westover KM and Bever JD. (2001). Mechanisms of plant species coexistence: complementary roles of rhizosphere bacteria and root fungal pathogens. Ecology, 82: 3285–3294.

White DJ, Haber E and Keddy C. (1993). Invasive plants of natural habitats in Canada: an integrated review of wetland and upland species and legislation governing their control. Canadian Wildlife Service, Ottawa, On. 121 pp.

White GM, Boshier DH and Powell W. (2002). Increased pollen flow counteracts fragmentation in a tropical dry forest: an example from *Swietenia humilis* Zuccarini. Proceedings of the NationalAcademy of Sciences of the USA, 99: 2038–2042.

Whitlock MC. (1999). Neutral additive variance in a metapopulation. Genetical Research, 74: 215–221.

Whitney KD and Gabler CA. (2008). Rapid evolution in introduced species, invasive traits and recipient communities: challenges for predicting invasive potential. Diversity and Distributions, 14: 569–580.

Wilcove DS, Rothstein D, Dubow J, Phillips A and Losos E. (1998). Quantifying threats to imperiled species in the United States.Bioscience, 48: 607–615.

Wilcox KL, Petrie SA, Maynard LA and Meyer SW. (2003). Historical distribution and abundance of *Phragmites australis* at Long Point, Lake Erie, Ontario. Journal of Great Lakes Research, 29:664-680. 78.

Williamson F and Fitter A. (1996). The characters of successful invaders. biological conservation,78:163–170.

Williamson M. (1996). Biological invasions. Chapman and Hall, London, UK.

Williamson M, Dehnen-Schmutz K, Kuhn I, Hill M, Klotz, S, Milbau, A, Stout J and Pysek P. (2009). The distribution of range sizes of native and alien plants in four European countries and the effects of residence time. Diversity and Distributions, 15: 158–166.

Williamson M. (2006). Explaining and predicting the success of invading species at different stages of invasion. Biological Invasions, 8: 1561–1568.

Willis JA, Groves RH and Ash JE. (1998). Interactions between plant competition and herbivory on the growth of *Hypericum* species: a comparison of glasshouse and field results. Australia Journal of Botany, 46: 707-776.

Wilson JB.and Agnew ADQ. (1992). Positive-feedback switches in plant communities. Advances in Ecological Research, 23: 263–336.

Wilson JRU, Richardson DM, Rouget M, Proches S, Amis MA, Henderson L and Thuiller W. (2007). Residence time and potentialrange: cructial considerations in modeling plant invasion. Diversity and Distribution, 13: 11-22.

Windham L and Lathrop RG. (1999). Efeects of Phragmites australis (Common reed) invasion on aboveground biomass and soil properties in brakish tidal marshes of the Mullica River, New Jersey, Estuaries, 22:927-935.

Windham L, Weis JS and Weis P (2003). Uptake and distribution of metals in two dominant salt marsh macrophytes, *Spartina alterniflora* (cordgrass) and *Phragmites australis* (common reed). Estuarine, Coastal and Shelf Science. 56: 63-72.

Witte F, Goldschmidt T, Wanink J *et al.* (1992). The destruction of an endemic species flock: quantitative data on the decline of the haplochromine cichlids of Lake Victoria. Environmental Biology of Fishes, 34: 1–28.

Wolfe BE and Klironomos JN. (2005). Breaking new ground: soil communities and exotic plant invasion. Bioscience 55: 477–487.

Wolfe BE, Rodgers VL, Stinson KA and Pringle A. (2008). The invasive plant *Alliaria petiolata* (garlic mustard) inhibits ectomycorrhizal fungi in its introduced range. Ecology, 96: 777-783.

Wright IJ, Reich PB, Westoby M, Ackerly DD, Baruch Z, Bongers F, Cavender- Bar J, Chapin T, Cornelissen JHC and Diemer M. (2004). The worldwide leaf economics spectrum. Nature (London), 428: 821–827.

Wright S. (1951). The genetical structure of populations. Annals of Eugenics, 15: 323– 354.

Wright S. (1931). Evolution in Mendelian populations. Genetics, 16: 97–159.

Yulan P, Weiguo T, Weikai B, Xinfen G, Ning W, Peng L and Weiyang X. (2008). Aboveground biomass allocation and growth of *Phragmites australis* ramet at four water depths in the Jiuzhaigou Nature Reserve, China. Journal of Applied Environmental Biology, 14: 153-157.

Zalapa JE, Brunet J and Guries RP. (2010). The extent of hybridization and its impact on the genetic diversity and population structure of an invasive tree, *Ulmus pumila* (Ulmaceae). Evolutionary Applications, 3:157-168.

Zedler JB and Kercher S. (2004). Cause and consequences of invasive plant in wetlands: opportunities, opportunists and outcomes. Critical Review in Plant Science, 23: 431-452.

Zenger KR, Richardson BJ and Vachot-Griffin AM. (2003). A rapid population expansion retains genetic diversity within European rabbits in Australia. Molecular Ecology, 12: 789–794.

Zhang DX and Hewitt GM. (2003). Nuclear DNA analyses in genetic studies of populations: practice, problems and prospects. Molecular Ecology, 12: 563–584.

Zhang Q, Yang R, Tang J, Yang H, Hu S and Chen X. (2010). Positive feedback between mycorrizal fungi and plants influences plant invasion success and reistance to invasion. PlOS 5(8): 12380.

Zhang YJ and Feng YL. (2004). The relationships between photosynthetic capacity and lamina mass per unit area, nitrogencontent and partitioning in seedlings of two Ficus species grown under different irradiance. Plant Physiology and Molecular Biology, 30: 269–276.

Zutshi DP. (1975). Association of macrophytic vegetation in Kashmir lakes. Vegetation, 30: 61-65.